A Polish Factory

A
POLISH
FACTORY

A Case Study of Workers' Participation in Decision Making

JIRI KOLAJA

GREENWOOD PRESS, PUBLISHERS
WESTPORT, CONNECTICUT

Library of Congress Cataloging in Publication Data

Kolaja, Jiri Thomas, 1919–
    A Polish factory.

    Reprint of the ed. published by University of
Kentucky Press, Lexington.
    Includes bibliographical references.
    1. Textile workers--Łódź, Poland. 2. Employees'
representation in management--Poland--Case studies.
I. Title.
[HD5658.T42P64 1973]   658.31'52   73-10736
ISBN 0-8371-7026-5

Originally published in 1960 by the University of
Kentucky Press, Lexington

Reprinted with the permission of the University
Press of Kentucky

Reprinted in 1973 by Greenwood Press,
a division of Williamhouse-Regency Inc.

Library of Congress Catalogue Card Number 73-10736

ISBN 0-8371-7026-5

Printed in the United States of America

Foreword

THIS BOOK represents a first in studies of industrial sociology. To my knowledge, this is the first time we have had a field research report on human relations in a factory in a Communist society.

Dr. Kolaja pioneered in securing research entry into the factory in the first place, and he has exploited this opportunity with skill and persistence. On the basis of interview data, we see how the workers regard their work, factory management, and the agencies theoretically designed to serve their interests. Perhaps of greatest interest in the book are Dr. Kolaja's observational reports of workers' council meetings. As we see worker representatives and factory management people grappling with or failing to come to grips with the problems they jointly face, we gain insight into the dynamics of organizational behavior in a Communist society.

Students of industrial sociology are displaying a growing interest in intercultural studies. For many years, we were basing our attempts at generalization on studies done primarily in the United States. With this limited perspective, we could never surely distinguish between behavior due to the particular organization of human relations within the plant studied and behavior that was culturally determined by the society in which the participants had grown up. The opening up of field studies in diverse societies enables us to sharpen our theoretical tools. As long as we have no knowledge regarding the way life actually is lived within Communist plants, our knowledge of the field remains seriously lacking. Dr. Kolaja has taken an important first step toward filling in this knowledge.

The book also provides some provocative ideas regarding the relationship among ideology, organization structure, and behavior. The Communist ideology assumes that in a Communist system there is no difference of interests between the workers and management which theoretically represents the workers. Therefore, there is no need to design organizations to represent the workers in their conflict of interest with management. As Dr. Kolaja amply demonstrates, the conflict of interest is there nevertheless. Since the ideology assumes the absence of conflict, we find that the actually existing organizations provide no help in resolving conflict.

WILLIAM FOOTE WHYTE

New York State School of Industrial
   & Labor Relations
Cornell University
January 15, 1960

Acknowledgments

IN THE SUMMER of 1957, Poland was experiencing a period of more liberal political control, which accounts for my good fortune in obtaining permission from the Polish Ministry of Light Industry to undertake research in industrial sociology in a factory in Lodz. I would like to express my appreciation to the Polish Academy of Science, Institute of Philosophy and Sociology, which sponsored my application to the Ministry of Light Industry. My thanks also are due to Dr. Józef Kądzielski, who assisted me in the research, and to other Polish sociologists in Lodz who gave me spontaneous help and advice.

The interpretation of data is my own, and neither Mr. Kądzielski nor anyone else in Poland should be held responsible for it. In fact, there was some disagreement in the definition of the problem of inquiry between my Polish colleague and myself during our cooperation in research. But this was natural because of the difference between the a priori dialectic Marxist framework on the one hand and the empirical structural-functional sociological approach on the other. Nevertheless, I hope that the research experience we both shared and the direct contact with the data will prove helpful to us in our further development.

I hope that this study also will contribute a little to the present endeavor of the Polish people to build for themselves a better society. My gratitude is directed also to the courageous

Polish workers, men and women, who were willing to answer our questions. The management of the plant should also be commended for its cooperation.

Since my childhood I have been much interested in Poland, its history, literature, and people. As my father was a lecturer in Polish language at the Masaryk University of Brno, Czechoslovakia, I had the opportunity to acquire knowledge of the Polish language and culture in my formative years. In 1946 and 1947, I visited Poland still as a Czechoslovak citizen. During the latter year, I was there for more than two months, helping the World Student Relief to distribute and direct help for Polish universities.

I was in Poland in 1957, as an American citizen. My studies in America naturally have had a great influence upon me. Professor William F. Whyte, of the School of Industrial and Labor Relations at Cornell University, had introduced me to the expanding new field of industrial sociology. His friendly encouragement has made it easier for me to complete this study, and my obligation to him is great.

Likewise my appreciation is expressed to Professors Temple Burling, Henry A. Landsberger, and Robin M. Williams, Jr., for their stimulating and helpful criticism and comments. Last, but not least, I also would like to express my appreciation to Dr. Martha Jane Gibson, a former colleague at Talladega College, for her assistance with my English, to Alfred O. Cabana of Talladega for his help in translating Polish textile industry terms into English, and to Dr. Harry Best of the Department of Sociology at the University of Kentucky, who read the book in galley proof. Finally, I am grateful to the University of Kentucky Research Fund for a grant which helped me to prepare the manuscript for publication.

I also would like to use this opportunity to remember my teacher the late Dr. In. A. Bláha, the first, and let us hope not the last, professor of sociology at Masaryk University of Brno.

Contents

FOREWORD, by William Foote Whyte   PAGE V

ACKNOWLEDGMENTS   vii

INTRODUCTION   xiii

CHAPTER 1. The Polish Workers' Council   1

CHAPTER 2. A Lodz Textile Factory   13

CHAPTER 3. Past Events in the Factory   39

CHAPTER 4. Eight Weeks in the Factory   57

CHAPTER 5. Analysis of the 13 Events   101

CHAPTER 6. Attitudes toward Production   116

CHAPTER 7. Interpretation and Conclusion   134

APPENDIX: Bylaws of the Workers' Council   147

INDEX   155

Figures

1. Indexes of Decision Sharing in Industry   PAGE XV

2. Layout of the "Corner" of the Weaving Shop   21

3. Structure of Organizations   27

4. Future Aspirations   37

5. Applications to Labor Union and Social Department during Two Years   54

6. Average Fulfillment of the Production Plan by Persons in Both Work Groups in Percentages   120

7. Causes of Low Production on Particular Days over Four Past Weeks   126

8. Perception of Work Performed the Previous Day, Reported for Six Days   127

9. Reasons for Bad Work Performance the Previous Day, Reported for Six Days   127

10. Causes of Production Troubles as Ranked by Management and as Ranked by Workers   129

11. Three Most Important Causes of Production Troubles Ranked by Management and Workers   130

Introduction

THIS STUDY of worker participation in the management of a
Polish factory, made from personal observation and from data
collected through interviews and questionnaires, provides
specific evidence both to correct and to support views now
prevalent not only in Communist countries, but also in the
West.

In broad, our study does lend support to recent criticisms
of traditional theory. Both Saint-Simon and Marx believed
that conflict between management and the worker would
vanish if their programs were put into effect, although for
different reasons. Saint-Simon, whom we may regard as an
exponent of the traditional view in western liberalism, assumed
that reason and science, once allowed their proper authority,
would be accepted by industrialist and laborer alike, and that
society would spontaneously be ordered into a natural and just
economic system. Marx, of course, felt that the same end
would be achieved, but only after the conflict between management and labor had been resolved by inevitable class struggle
ending in the triumph of the proletariat; as the capitalist class
vanished and the worker assumed ownership of the economy,
he would naturally approach his work with both the interest
of the owner and the satisfaction of the laborer over worthwhile achievement.

After studying Polish efforts at more satisfactory management-labor relations, however, we would be inclined to hold that not only Saint-Simon but Marx, too, was overoptimistic about the prospects for the disappearance of this conflict. The Polish worker, like others, is apparently more affected by the problems common to modern industrialization than by economic and political philosophies in his attempts to solve his problems of industrial relations. The Polish worker, like his fellow in the economies of the West, is plagued by the monotony common to workers in industries engaged in mass production; he shares with his western fellow laborers a concern for his own personal affairs and frequently a lack of interest in the managerial and organizational problems within his factory (even though the industry is state-owned); he is continually seeking more meaning and enjoyment of life outside working hours; and his attitudes reflect his cultural and social differences, as do the attitudes, say, of American and British workers.

Those who deny that Saint-Simon and Marx were overoptimistic about the prospects of achieving an efficient and harmonious society generally look to our changed social and economic conditions to provide an explanation for a century of not always encouraging experiences. Both Marx and the liberal reformers lived in an age of face-to-face relationships in relatively small industries. Robert Owen, for instance, developed designs for industrial villages of just 500 to 1,500 persons. The modern divorce of the owner from the control and management of the means of production and the growth of modern industries to their present huge proportions have created problems of impersonal relationships and of communication in presenting the complexities of the modern industrial plant to the worker in terms which he can readily comprehend.

The view generally held in America is that "free men work best." Studies have (with important exceptions) tended to support the contention that workers' participation in management, developed in a democratic way, has increased both plant productivity and workers' satisfaction.[1] There are, of course, different degrees of participation, ranging from simple sharing of information to what Shuchman calls codetermination in policy decisions.[2] Figure I, employing the different types of participation developed in a study by the International Confederation of Free Trade Unions, lists the stages of participation from none at all to full worker control[3] and relates them to four gradually increasing degrees of participation.

FIGURE I

INDEXES OF DECISION SHARING IN INDUSTRY

EXTENSITY	INTENSITY			
	Reception of Information	Presentation of Opinions	Rejection of Demands	Ability to Move Others
No Consultation	1-1	1-2	1-3	1-4
Safety	2-1	2-2	2-3	2-4
Social and Sport	3-1	3-2	3-3	3-4
Production Activities	4-1	4-2	4-3	4-4
Personnel Counseling	5-1	5-2	5-3	5-4
Working Conditions	6-1	6-2	6-3	6-4
Auditing	7-1	7-2	7-3	7-4
Profit Sharing	8-1	8-2	8-3	8-4
Hiring and Firing	9-1	9-2	9-3	9-4
Management	10-1	10-2	10-3	10-4
Full Workers' Control	11-1	11-2	11-3	11-4

Not all studies agree with the "free men work best" concept, however. Whyte and Miller have pointed out that in most of these studies the role of the leader was not fully spelled out in the majority of experiments with the "democratic" process of decision sharing. Did the "boss" really accept the group

[1] For a survey of pertinent studies undertaken by Kurt Lewin, Lester Coch and John R. P. French, Jacob Levine, and John Butler, Nathaniel Cantor, Eugene H. Jacobson, Nicholas Babchuk and William J. Goode, A. K. Rice, Jay M. Jackson and others, see Frederick Herzberg and others, *Job Attitudes: Review of Research and Opinion* (Pittsburgh, Psychological Service of Pittsburgh, 1957), 123-66.

[2] Abraham Shuchman, *Codetermination, Labor's Middle Way in Germany* (Washington, D.C., Public Affairs Press, 1957), 6.

[3] International Confederation of Free Trade Unions, *Workers' Participation in Industry* (*Study Guide No. 3*, Brussels, 1956), 16.

decision, or did he only consult the workers and skillfully persuade them to accept his decision?[4] While most studies agree that worker participation in decisions increases worker satisfaction through self-expression, there are in fact many situations where the issues to be decided involve such large groups, such complex questions, and such great distances between groups, as with unions, that the sense of participation is not always real to the individual worker, for a worker's satisfaction and continued participation in decision sharing is likely to be dependent upon the likelihood of having his suggestions accepted. Further, when we say "free men work best," we must concede that they do not always produce high-quality work as a group; work requiring a high degree of concentration, such as that of the artist, author, or mathematician, will actually suffer from interaction, at least during certain periods, with other persons. Some studies, indeed, have shown that productivity can be high when both low group morale and hostility toward management are present; others indicate that the introduction of sometimes cumbersome techniques for group sharing in decisions may actually retard plant efficiency.[5] Furthermore, studies have shown that the most frequent way of helping the worker to identify himself with the interest of the enterprise—profit sharing—seems to be more important for the building of morale than for output itself.[6]

[4] William F. Whyte and Frank B. Miller, "Industrial Sociology," in Joseph B. Gittler (ed.), *Review of Sociology: Analysis of a Decade* (New York, John Wiley, 1957), 307.

[5] William J. Goode and Irving Fowler, "Incentive Factors in a Low Morale Plant," *American Sociological Review*, XIV (October, 1940), 618-24; D.G. Marquis and others, "A Social Psychological Study of the Decision-Making Conference," in Harold S. Guetzkow (ed.), *Groups, Leadership and Men: Research in Human Relations. Reports on Research Sponsored by the Human Relations and Morale Branch of the Office of Naval Research, 1945-1950* (Pittsburgh, Carnegie Press, 1951), 56-67; Alex Bavelas, "Communication Patterns in Task-Oriented Groups," in Daniel Lerner and Harold D. Lasswell (eds.), *The Policy Sciences: Recent Development in Scope and Method* (Stanford, California, Stanford University Press, 1951), 193-202.

[6] See a survey of 600 British firms, referred to by W. Robson-Brown and N. A. Howell-Everson, *Industrial Democracy at Work: A Factual Survey* (London, Sir Isaac Pitman, 1950), 13. A similar conclusion is reached by Fred H. Blum, *Toward a Democratic Work Process: The Hormel-Packinghouse Workers' Experiment* (New York, Harper, 1953), 72.

So, while psychological and sociological studies have corroborated the popular notion that "free men work best," they have introduced a number of important qualifications to the thesis. The cultural values in America and Great Britain favoring the democratic tradition seem to have some influence both upon the attitude toward the "free men" idea and upon its success once adopted.

It has been shown that the general type of worker participation is influenced by cultural, social, and personal factors. For example, comparison of British and German approaches to codecision reveals the German tendency toward a deductive development of universal systems as reflected in the comprehensive German codetermination law of 1951. British empiricism, on the other hand, is reflected in the more variable British labor participation patterns.[7]

Social situations also influence the nature of participation. In the cited study by Marquis and others it has been suggested that goal-directed, busy men in conferences are more concerned with achievement than with the maintenance of the group. Again, an emergency situation likewise may produce leaders who rely but little on group participation. Industry generally, however, is characterized by order, repetitive routine, and long-range planning of future needs. This type of work group, as contrasted, say, with a military force, is better suited for group discussion and group decisions.

Naturally, the personalities, both of the individual worker and of members of management, will have much to do with the determination of the degree of decision sharing by each group. In our Polish factory, the domineering character of the plant director tended to reduce the amount of workers' participation, while among the workers, the presence of an engaging and forceful personality, like "I-5" in the weaving

[7] See Herbert J. Spiro, *The Politics of German Codetermination* (Cambridge, Massachusetts, Harvard University Press, 1958), 10-14; Jeremiah Newman, *Co-responsibility in Industry: Social Justice in Labour Management Relations* (Westminster, Newman Press, 1952), 67. See also James G. Abegglen, *The Japanese Factory: Aspects of Its Social Organization* (Glencoe, Illinois, Free Press, 1958), 116-32, for worker participation in another culture.

department, might increase the share of workers' participation.

Our study of worker participation in management in a Polish factory will, we hope, provide students of this important question with certain insights of unique value. For with its western cultural ties, coupled with its present involvement in the Soviet political and economic system, Poland's situation may furnish us with clues to the effectiveness of Communist theory in improving worker happiness and efficiency in particular, and to the problem of workers' participation in management in general.

1

The Polish Workers' Council

IN THE FIELD of industrial labor relations, Poland went through approximately the same stages as did Western European countries, although industrialization and other concomitant changes appeared somewhat later and the Polish labor movement had its own political and national character.[1] The Polish Socialistic Party (PPS), one of whose leading members for some time was Marshal Józef Piłsudski, was one of the most active forces helping to bring about the restoration of political independence in 1918. Trained in illegal and direct activities against the Czarist authorities, Polish organized labor entered the new Polish state proud and self-conscious. Compared with the peasant majority, industrial labor was socially and culturally more advanced.[2] Trade unions during the twenty years of the Polish Republic were governed by the Central Committee of Labor Unions (CRZZ), which was organized on an industrial basis, covering practically all trades and industries with the exception of the Union of Printers. Outstanding, through its social and educational activities, was the Union of Railway

Workers (ZZK), which had its own clubhouses, orchestra, vacation boardinghouses, and publishing firms.[3] Altogether, it is estimated that some 25 percent of Polish labor belonged to unions.[4] If only manual laborers are considered, the 1,200,000 union members constituted more than half of the total manual work force in Poland.[5]

During the twenty years of political independence, the cooperative movement, represented especially by the organization called *Spolem*, was expanding considerably. At the same time, schools were founded for workers, the so-called TUR, Peoples' University.[6]

Immediately after the World War II, the unions officially resumed their activities, returning to prewar labor law and administration of labor relations.[7] At the First Congress of Labor Unions in 1945, the unions declared themselves independent of political parties and of the government. This was possible because the Polish Socialistic Party was still independent and competing for the allegiance of workers with the Polish Communist Party (PPR). With the forced merger of the PPS with the PPR in 1949, the former relative independence of the labor unions was lost.[8] In 1949 the Second Congress of

[1] Aleksander Wóycicki, *Dzieje Robotników Przemysłowych w Polsce* (Warsaw, F. Hoesick, 1929).

[2] Alicja Iwańska (ed.), *Contemporary Poland: Society, Politics, Economy* (New Haven, Connecticut, Human Relations Area Files, 1955), 113. In 1930, about 60 percent of the population depended upon an income earned in agriculture, while in 1950, the same category constituted no more than 47 percent. See Główny Urząd Statystyczny Polskiej Rzeczypospolitej Ludowej, *Rocznik Statystyczny, 1957* (Warsaw, 1957), 21.

[3] Feliks Gross, *The Polish Worker: A Study of a Social Stratum* (New York, Roy, 1945), 148.

[4] Clifford R. Barnett and others, *Poland, Its People, Its Society, Its Culture* (New York, Grove Press, 1958), 193.

[5] Celia Stopnićka Rosenthal, "The Functions of Polish Trade Unions: Their Progression Toward the Soviet Pattern," *British Journal of Sociology*, VI (September, 1955), 265.

[6] Gross, 152, 250-51.

[7] Kazimierz Grzybowski, "The Evolution of the Polish Labor Law, 1945-1955," in Z. Nagórski (ed.), *Legal Problems Under Soviet Domination* (Association of Polish Lawyers in Exile in the United States, *Studies*, I, New York, 1956), 81.

[8] The new party was called the "Polish United Workers' Party" (PZPR). In further references, the word "party" will mean the PZPR.

the Labor Unions accepted the principle of "democratic centralism." The function of the labor unions, as laid down in the 1949 statutes, was to support government policies in the economic field by mobilizing the working masses to fulfill production requirements ahead of time, to organize socialist competitions, and to enforce labor discipline.[9]

Unions, having become "the basic transmission belt of the party to the non-party masses,"[10] went along with the passing of the severe Labor Discipline Law in 1950. By this year, the party had gained control over unions. The law gave wide disciplinary powers to management and provided for court trial of serious cases of infraction. As little as one day of unjustified absence was cause for disciplinary action by deduction of wages, reprimand, or transfer to a lower paid job. An absence of four days called for court action, and the worker could be sentenced to a term of correctional labor, receiving wages reduced by 10 to 25 percent.[11]

Collective bargaining had become in effect a collective agreement to tie earnings to the fulfillment of production targets set up by higher authorities. The following points were included in an agreement: organization of socialist competition, organization of a Stakhanovite movement, extension of the piecework system, economy in the use of materials, fuel, and similar goods, and support of workers' inventiveness for labor and material cost-saving devices. No bargaining in respect to wages was admitted. The wages were defined in the technical standards of output established by the competent ministry after consultation with the national board of the labor union.[12]

Another important change was a governmental decree on October 26, 1950, which introduced "the principle of individual management and personal responsibility of the manager, relieving him of all interference by the labor union."[13] Thus, the labor union in effect had become a social department of management, whose function was to stimulate workers, to increase production, and to take care of some welfare activities.

[9] Grzybowski, 84.   [10] Rosenthal, 265.   [11] Grzybowski, 92.
[12] Grzybowski, 87.   [13] Grzybowski, 85.

A trend toward liberalization had already been in progress in Poland before October, 1956, when Władyslaw Gomułka returned to political power. The Labor Discipline Law itself was amended in 1955 by a decree of the Council of State and was abolished a month before the "Polish October Revolution" in 1956. During the earlier Stalin period, other matters were gradually taken out of the workers' hands, but beginning with 1954, this trend was reversed. For example, in 1950, special labor courts were abolished and cases were brought either under the regular courts or, in some instances, under the Enterprise Arbitration Boards, whose members were nominated by the government. In 1956 the boards were renamed "commissions" and their labor representatives were elected by the workers.[14]

Immediately after the liberation in 1945, shop committees (*rada oddziałów*) spontaneously organized by workers appeared in several factories with the purpose to bring back into production the industrial establishments in the heavily damaged and exhausted country. The government encouraged workers' participation in management by issuing a decree of February 6, 1945, stating that a shop committee representative was expected to sit on the managerial board of industrial enterprises. The law provided for election of a shop committee in every establishment with more than twenty workers. Another decree of January 3, 1947, confirmed the workers' participation in the management of the enterprise.[15]

During 1949, with the forced merger of the Polish Socialistic Party with the Polish Communist Party, the shop committees became subordinated to labor unions "with the effect that the shop committees, although elected by the factory crews (whether members of the union or not) were made into trade union local organizations, and subordinated to a complex hierarchy of labor union administration with the Central Council of Labor Unions as the highest body." The party by that

[14] Barnett, 197, 200.
[15] Kazimierz Grzybowski, "Polish Workers' Councils," *Journal of Central European Affairs*, XVII (October, 1957), 274.

time had obtained control of the labor union; the shop committee laws were never formally repealed but in practice made ineffective by the introduction of new political controls.[16]

After 1953, the first labor law reforms were primarily intended, Grzybowski says, to "stem the rising tide of discontent." In August, 1956, an effort was made by the Central Council of Labor Unions to draft a new law on shop committees. Despite provisions for workers' participation in disposing of the enterprise fund for the benefit of workers, in working out the production plan, in determining the use of investment appropriations, and in deciding on personnel problems, the proposed shop committee was designed to be nothing but a lower echelon of the labor union.[17] The labor union, however, through its former subordination to management and the party, was already discredited in the eyes of the workers. During the dramatic events of the Poznan workers' uprising and the later October events that swept Gomułka into power, labor unions did not play any significant role.

In October, 1956, the scene was cleared for the creation of the workers' council (rada robotnicza), a new independent institution related to the shop committees of the postwar period. In some factories the workers themselves again took the initiative, as in the years of 1944-1945, and started the councils ahead of the official legalization.[18] Indeed, the workers' councils of 1956 can be considered as continuations of the 1946 shop committees.[19] By April, 1958, about 5,600 enter-

[16] Grzybowski, 275. See also Barnett, 195.

[17] Grzybowski, "Polish Workers' Councils," 277.

[18] Roman Fidelski, O Radach Robotniczych: Nowe Zjawiska w Życiu Gospodarczym Polski (Warsay, Książka i Wiedza, 1956), 9-10.

[19] Three authors living in Poland today have referred to the 1945 shop committees as "workers' councils." "In Poland the workers' councils played a major role in the years 1944-45 when they took over factories and put them into operation often long before the decree was issued on the nationalisation of industry, and preceding a decision by central bodies then at the stage of being organized. Nevertheless, those councils were an institution characteristic for a period of transition. The stabilization and the organization of the central authorities in most cases removed them altogether." Halina Białek and others, "The Workers' Council in Poland" (International Sociological Society, Industrial Sociology Section, Working Paper 8, mimeographed, 1957), 1.

prises, including almost all major production units, had their own workers' councils.[20]

The workers' council, though influenced by the similar Yugoslav organization, can be considered as a unique Polish economic institution. Evoking great hopes, it has become a symbol of the desovietization of Poland since October, 1956, an example of the "Polish road toward socialism."

The bill establishing the workers' councils was passed November 19, 1956. One provision made workers' councils mandatory in state enterprises if the majority of employees favored them. The major function was to increase workers' participation in the management. Among twenty-one sections listed by the bill, that most relevant to our analysis was Section 2-2: The workers' council acts on the basis of production goals set by the national economic plan. It strives to develop the enterprise, to increase production, to decrease the costs of production, and to improve the quality of products, while simultaneously striving for improvement of work conditions and welfare of the employees.[21] Because this section directed the workers' council to watch not only production, but also the welfare of workers, its relation to the labor union was not quite clear.[22] However, several analyses have stressed that "the labor union will resume its traditional task of defending workers' interests as employees, while the workers' council will represent the role of workers as owners and managers of the plant."[23]

Other provisions of the bill provide: that the workers' council ought to be composed, as far as possible, of at least two-thirds of the workers (Section 7-2); that the chairman or the vice chairman cannot be the plant director or the vice director

[20] See the figure given by Gomułka during his speech at the Fourth Congress of the Labor Unions in Warsaw. Warsaw *Trybuna Ludu*, April 15, 1958.

[21] See Fidelski, 56.

[22] A conflict between the labor union and the workers' council was reported by Wilhelmina Skulska, "Uwaga! Rady Górnicze Zagrożone," Warsaw *Po Prostu*, June 30, 1957.

[23] Białek, 5.

(Section 8-2); that every employee is assured free access to meetings of the workers' council (Section 10); that the plant director and the vice director are nominated or relieved by the appropriate governmental authority, but the nomination must be approved by the workers' council (Section 12).

The director of the plant could veto measures passed by the council, and in case of disagreement, both parties could bring the matter to higher authorities. The director was responsible for the execution of the workers' council program; he was the highest authority on personnel problems. However, he could not remove an employee from his position as long as the employee was serving his term as a member of the workers' council. The decree also provided for freedom in developing local rules for the election of representatives, and the formal rules under which the work of the council ought to be carried out. It was also stated that membership in the council ought not to bring about a change in the worker's job. It was emphasized that in such a way one ought to keep alive the identification and communication of the rank and file with the council.[24]

The director had two major responsibilities. He was responsible to the state, which nominated him and in whose name he administered the factory, and also to the workers' council, whose decisions he was supposed to implement, provided that an agreement between the council and the director was previously reached. Actually, these two roles of the director were a hidden source of potential conflict between the state and the workers. In Yugoslavia the conflict was theoretically reduced by defining the means of production as social, not state, property. Thus: "In the view of theoreticians, the relationship proper to socialism could not be established if the means of production continued as state property. These had to be made the property of society, and be placed under the control of representative bodies. . . . Concrete rights of administration and utilization were to belong to producers

[24] Fidelski, 15.

themselves."[25] Nothing like that was conceived in Poland. The workers, in all definitions of their role, were entitled merely to comanagement (współzarządzenie), that is, to supervise and control the administration of the enterprise, as defined by Gomułka.[26]

The second difference from the Yugoslav workers' councils lay in their relationship to higher national bodies.[27] The Polish groups were not related to any higher assembly, but operated individually only within the particular factory, being responsible only to the crew of the factory which elected them. Thus, the central state and party authorities could influence the workers' councils only through their local party, management, and labor union people. The Polish workers' councils were then a mixture of dependence upon and independence of the state. The Polish economic system was still centrally planned and led, a sort of intermediate step between the Soviet system and the Yugoslav socialistic economy.

To satisfy the desire for some degree of entrepreneurial freedom and to save material, a decree (No. 704) permitting some small independent production by the enterprise accompanied the law on the workers' councils. The newly constituted workers' council could thus become engaged in so-called "sideline" activity, the products of which could be sold by the plant itself. Furthermore, the plant was given a free hand in other respects—the disposal of used machinery, for example.

Another bill, issued November 19, 1956, introduced a so-

[25] For the Yugoslav Workers Management Law, issued July 2, 1950, and other pertinent decrees, see Alexander Adamovitch, "Industrial Management in Yugoslavia," *Highlights of Current Legislation and Activities in Mid-Europe,* V (March, April, 1957), 165-78. The quotation is from p. 170.

[26] Janina Miedzińska, "Kryzys Samorządu Robotniczego," *Kultura,* No. 10/120 (October, 1957), 78. Concrete examples of what the workers' council did or could do are to be found especially in Chapter IV and the Appendix. It should be pointed out that most persons to whom we talked were more or less uncertain what the role of the workers' council ought to be. In this sense, the workers' council was really "an experimental institution."

[27] Alexander Adamovitch, "Contemporary Yugoslav Trade Unions; I. Phase (January 3, 1945–October 29)," *Highlights of Current Legislation and Activities in Mid-Europe,* IV (October, 1956), 326.

called "plant fund," in order to motivate labor to become more involved in greater efficiency of production (Section 1). A part of the total profit was to constitute the fund, which could be used either for individual premiums or for collective goals such as construction of new housing facilities. Section 5 also stated that the maximum limit to which the fund could be raised would not be more than 8.5 percent of the total wages and salaries paid by the plant during one year; that is, approximately one-twelfth of the annual pay. Therefore, it was popularly called "the 13th wage."[28]

As in a capitalistic country, where shareholders control the director, here the workers' council should assume the same role. However, some authors have found that the workers' councils are largely dependent in their decisions upon factors which are beyond their control or which are a result of chance.[29]

It was our privilege and good fortune to be able to undertake a study of one of these workers' councils, one that probably was neither the best nor the worst.

From July 11 to September 3, 1957—a total of forty-two working days or nights—under a permit from the Polish Ministry of Light Industry, I studied productivity in a textile factory. My Polish colleague and I selected two groups of workers with significantly different productivity, hoping that by controlling the groups on as many factors as possible, we might discover data to account for the work achievement differences.

While we were engaged in collecting observational, attitudinal, and other information on the two groups, we came to realize that productivity differences between them were in

[28] Fidelski, 47; Janina Zajdzikowska and Teresa Czacka-Nowakowska, "Organizacyjne i Gospodarcze Problemy Samorządu Robotniczego," Ekonomika i Organizacja Pracy, VIII (February, 1957), 53.

[29] Wiesław Krencik and Czesław Niewadzki, "O Właściwą Rolę Rad Robotniczych w Polskim Modelu Gospodarczym," Gospodarka Planova, XII (March, 1957), 5; Władysław Piotrkowski, "Samorząd Robotniczy a Zarządzanie Przedsiębiorstwem," Zeszyty Naukowe Wyższej Szkoły Ekonomicznej w Łodzi, No. 3 (1957), 43.

fact not significant. I realized that organizations existing on the factorywide basis were of greater relevance to the problem of productivity than small group factors. These plant agencies were relatively new and had not yet been generally accepted, and their influence transcended that of small groups within the factory.

Thus, gradually we shifted our attention from the two small groups to the plantwide organizations. Fortunately, some of the data on the two small work groups could be related to our exploration of the functions of the workers' council and other agencies.

In addition to the workers' council, the other organizations within the factory that influenced, or sought to influence, workers in their productive efforts were the management, the labor union, and the party. Interestingly, several events provided us also with details of spontaneous collective behavior of a group of workers that was carried on outside the framework of the four organized bodies.

From the methodological viewpoint, then, this analysis is primarily a case study of a workers' council in a factory. How was the council developed; how did it function? We asked respondents in an unstructured way to give us an account of the events before our arrival, and we attended some interesting sessions of the workers' council during our two months stay in the factory.

One type of data was "historical" (either reported by participants or directly observed by us) and was later, either immediately or in the evening, recorded in a diary. In addition, data were collected that can be characterized as "cross-sectional." These were attitudes, as expressed through answers to short questionnaires not amounting to more than ten questions each, degrees of knowledge disclosed by the familiarity with names of supervisors and organizational rules, direct observation of frequency and direction of interaction between persons in the group studied, choices concerning the causes of troubles in the production process, data from personnel department records, and production and earnings data for a period over six months preceding our arrival as well as the same data for the period of our attendance. The sources of data and the instruments and conditions of administration will be described in following chapters. It should also be pointed out that merely the most relevant data will be presented in this report. Actually, we have collected almost twice as much data as will be reported here.

Most of the questionnaire data were collected by means of short schedules that were administered during working hours, either by asking a worker to answer a few questions while he was standing at the loom, or by inviting him to a small room —"the smoking room"—that was located in the immediate neighborhood of the weaving shop (see Figure II). This explains why the schedules had to be short. Furthermore, since we were continuously involved in a flow of events, which provided us with new ideas, we preferred to keep our research design in a rather adjustable form. For example, complaints by our respondents concerning the technological troubles in their work, answered by public pronouncements by the plant director that the technological factors were under control, led us to the development of a multiple-card questionnaire.

The shop itself and persons in it were not the sole source of data. As soon as I had become accepted by the workers and management personnel, I met them also outside the factory, frequently being invited to their homes. Through direct observation and through additional schedule data, some background information on living conditions of the respondents was obtained. These data explain, for example, why there was such a great desire for "a new apartment" (in most cases meaning one room).

Now, as far as the definition of our research problem was concerned, the question of the major function of the whole organization was easily and simply answered: it was supposed to produce as much as possible with as little expenditure as possible of energy and raw material. All persons presumably were agreed upon this rational and economic definition of the major purpose—but were they really?

Secondly, if those who worked in the factory did not perceive their individual functions as in accordance with the major productive-economic function, what were the differences in attitudes and how could they be explained?

Third, what programs and activities by persons, organizations, and institutions furthered the major productive-economic function, and what hampered it? What was the role of the workers' council?

Fourth, if there were difficulties in the performance of the major function, what could best bring about an improvement?

The four questions referred principally to the tasks and problems assigned to the workers' councils.

As one could expect, my role as a researcher from America was unusual and produced some distrust. Under Stalin, as Poles would say, nobody but the secret police had ever interviewed the people in the shop. So far, they had never been exposed to a sociological inquiry. Also, my United States passport and my Czechoslovak accent in my Polish speech contributed to general surprise and wonder as to my identity. However, it should be stressed that I was not the first American citizen whom they had seen. A few months before my arrival, an American newsmen's group had passed along their looms.

Despite my explanation of my purpose in studying them at their looms and visiting them in their homes, there was some misconception of my identity and my work, which I combated by writing a personal letter to each of them.

The roles ascribed to me at first were "good" or "neutral," such as "American," "Professor," or "Czech," as well as "bad," such as "Western Spy," "American Communist," or "Management's Ally." My Polish colleague mediated and helped to remove these misunderstandings, and on the whole, respondents were in most cases eager to answer most of our questions. They felt gratified that somebody came down to their looms and worried about their problems and work.

2

A Lodz Textile Factory

THE CITY OF LODZ, the second largest of contemporary Poland, has a population of close to 700,000. Its fame, based on its textile industry, especially the cotton and wool branches, has given it the title, the Polish Manchester. Before the rebirth of the Polish state in 1918, Lodz (being located in the territory of so-called "Kongresówka") belonged to Russia. Opening of a vast Russian market gave a mighty stimulus to the industry.

The factory we studied concentrated on the production of cheap cotton material. By 1910, the plant employed some 7,200 workers;[1] during our visit in 1957, employees numbered about 8,000. Its peak of expansion had already been reached before World War I. Since then, additions to the plant were hardly noteworthy. Most machines had been installed before 1910, and the buildings, with the exception of two additional floors added to the major building, seemed to be of pre-1920 architecture. When in World War II, Lodz and its surrounding villages were incorporated into Germany, the factory was run by Germans, who held most of the supervisory positions. Since

1945, the plant has become the property of the Polish state, being managed by state-appointed directors.

Twelve years of state management did not introduce many technological changes except for a few post-World War II machines of Swiss and Polish make. What was new on the grounds of the factory was a medical checkup center and a home for children whose mothers work in the factory.

The city of Lodz is not beautiful; being located on a plain and without any major river, the city does not have any dominant landmark. Nor is there any significant architecture or historic place that would harbor national or religious symbols of the Polish people.

The factory is surrounded by rather shabby red-brick houses and fields. Earlier, at one side of the factory block there was a Jewish part of Lodz, partially destroyed by Nazis. On the whole, a newcomer from the West is depressed by the neighborhood.

The ugly old red-brick apartment houses opposite the factory, built by the former factory owner, are occupied by the families of persons who have worked in the plant for most of their lives. Due to the great housing shortages, new workers would not get an apartment here. In the vicinity of these apartment houses is a small chapel, surrounded by a few trees. This is the only place where a person might enjoy sitting outdoors. Here I met several individual workers for informal interviews.

The factory, also a red-brick four-story structure, protected by a high brick, iron-barred wall, looks like a prison. This unfavorable impression is further brought out by the bodily search of workers as they leave the factory. The main entry is divided into a corridor for women and one for men. As employees walk out, special guards check to see whether they are carrying out pieces of material under their outer garments. Fortunately, an entry card exempted my Polish colleague and myself from such a search. Another, smaller entry leading

[1] Wóycicki, *Dzieje Robotników Przemysłowych w Polsce*, 195.

directly into the office building, used mostly by the management people, consists of a villa built by the former owner at the beginning of the century. Its ornaments and functionless shape contrast strangely with the utilitarian architecture of other factory buildings. It was here that my Polish colleague and I entered the factory for the first time.

The factory block comprises eight closely grouped buildings. The major four-story building contains the spinning department; the weaving department occupies three other buildings. The finishing department, where the textile material is dyed and prepared for distribution, has its own separate structure. In addition to these three production departments, there is the maintenance department, composed of several units such as machine and electrical shops. Behind this group of buildings there is a large free area, open for further development. Here, standing apart from the production structures, is a new two-story structure, the medical checkup center, constructed after World War II. The home for children, also a post-World War II construction, is located outside the factory block, facing the entry to the administration building. Finally, it should also be pointed out that the cafeteria for workers is in one of the production structures. Another post-World War II innovation, the factory broadcasting center, the trade union offices, and the party offices are located directly in the central production building.

On the whole, the organization of the area is purely functional. Only at the major entry is there a small flowerbed. With the exception of the cafeteria, there is no rest or recreation area. One could often see people waiting for their shift change standing or sitting on the ground with their backs against the wall of the buildings.

Having obtained the necessary permission from the Polish Ministry of Light Industry in Warsaw, my Polish colleague and I were introduced by a secretary to the office of the plant director. The plant director was not in; we talked, however, with the director of administration, who came to join us in the office.[2] The most spectacular feature of the room was four red flags carrying symbols of the textile industry. These were the honor flags, production trophies won by the factory in competition with other plants.

The management had already been informed of our coming by a letter from the ministry in Warsaw, and was consequently most cooperative. The chairman of the workers' council presidium joined us in our discussion. Later, the plant director himself came in and listened for a short while to our conversation, answering meanwhile a few telephone calls and signing some papers brought in by the secretary.

We noted that all men wore grey or light blue coveralls, as did the women, also. The men did not wear ties.

We were taken around the factory by one of the younger supervisors, and the next day we were introduced to the manager of the weaving department. We found the manager, a young man, to be most cordial. After a few introductory statements, I was asked about the living conditions of American workers; it was of interest to me that the top management did not show any curiosity of this kind.

From the weaving department manager we went down in order of rank to the shift supervisor, and from him to the section foreman, and from the latter to the foremen, who brought us directly to the operating looms.[3]

Compared to the recommended "double entry" in industrial studies, our approach could be criticized, for we had entered the industrial organization via management.[4] However, the labor union was relatively unimportant, and workers did not feel that they had any organization that they could consider their own. Thus, we could not have used another "entry" because there was no organization comparable to those found within the American industrial power structure.

2 An approximately corresponding position within an American plant of a similar size would be that of the assistant plant manager.

3 Though the Polish foreman is technologically performing almost the same job as the American loom fixer, he has traditionally a higher status.

4 See Robert Kahn and Floyd Mann, "Developing Research Partnerships," *Journal of Social Issues*, VIII, No. 3 (1952), 4-10.

In the five days we were exploring possible groups to study, we paid visits to the party headquarters, to the labor union, and to workers' council organizations, whose good will we could assume, since, as will be shown later, they were parallel and complementary to the management.

We were introduced to the two groups of workers we had selected first by being introduced to the foremen of both work groups by the shift supervisor and the section foreman. It happened that the electric current went out of order and, therefore, the foremen could call their people together and introduce me. I immediately became an object of great attention. Those who stood around me, mostly women, started to express their complaints to me. I told them that I would like to stay with them for several weeks, coming to the shop and observing and talking to them. One of the women said: "Life is getting worse and worse." Somebody else shouted: "They swindle us; they take care only of themselves." "There have been so many different deductions from our paychecks, that I do not remember what it was about," said one woman, answering a question on how much she earned. "They always figure out something," she added. "I earn only enough to live," stated another voice.

The situation was unusual. There I was standing among men and women dressed in their grey work clothes, some of them sweating in the artificially damp atmosphere of the weaving shop, all of them shouting out their complaints and disrespect and suspicions against the management. One older woman said to me: "You see, this is the life of the proletariat," and raised her hand pointing to the glass roof from which water was leaking down on the looms. Jokes about the bellies of the management people were heard. The crowd around me was stimulating and encouraging itself. As the situation seemed to be developing into a demonstration, I was glad that my coworker was coming to join me, accompanied by the section foreman. The section foreman, a tall and neatly dressed man, quieted the people down. "Well," he said, "you have only complaints." "And how about you?" some persons shouted at him, and the group started to laugh. Along with their outspoken bitterness, there was an obvious lack of fear, a sort of cheerful sneering at management. I later discovered that the men and women thought I was there just for a while; therefore, they were trying to communicate to me their dissatisfaction in a hurry—exaggerated and dramatized by crowd interstimulation.

At this first meeting, the rank and file stressed three sources of their dissatisfaction: first, the low pay; second, bad working conditions, such as heat and dripping water; third, their distrust and disrespect for the management. We shall see later that the financial theme, the ecological-technological theme, and the relationship with management were most frequently listed as sources of dissatisfaction of the men and women studied. It is also of interest to note that while workers spoke their mind at the first meeting in a very open manner, they later passed through periods of trust and distrust toward me. Apparently under the pressure of the group stimulation and anonymity (because I did not yet know anybody in the group), they aired their problems in an uninhibited way that I could never get them to do later.

Our original research design was the study of productivity of two small groups. After a five-day exploration of different departments and their work groups, we selected the weaving department and within it two groups on the same shift, working side by side. The groups were also equated as much as possible on the average age, sex ratio, and length of work within the group. The foremen of each of the two groups were said to have different personalities, while the groups were reported to have different productivity.

During the research our attention was gradually shifted more and more from the two small groups to the workers' council and other organizations in the factory. The reason for this shift was: (1) a tentative analysis of the data disclosing that production differential between both groups was quite small despite the information given us earlier by the person in charge of statistics in the weaving department; (2) our increasing realization that the organizations were more important for production because of their all-embracing influence and innovative character; (3) our inability to control satisfactorily all factors accounting for greater or smaller productivity of either group.

Although data were in many instances collected separately for each group and so differentiated in some tables, we came to treat them as one group of men and women in regard to organizations in the factory. Therefore, where it was feasible, the data were combined.

Why did we select the weaving department?[5] My Polish colleague was formerly a textile worker himself, a weaver in the plant some eight years earlier. Both his parents had worked in the factory; his mother was still living in one of the red-brick apartment houses. Therefore, it was to our advantage to settle at this plant and in this department because of my friend's familiarity with the job and with the people.

Another reason was that the chairman of the workers' council reported this department to be more articulate in its demands than any other part of the plant. The management also felt that this department was causing them the most problems. As we were interested in problem situations, we welcomed the characterization. Finally, in a more or less accidental way, we discovered also that a worker who had been elected to the workers' council presidium was in our selected group. Ultimately he proved very valuable for our understanding of the operations and problems connected with the new (at our arrival, just half a year old) organization.

The weaving shop was relatively large, crowded with some 1,200 looms; this shop received its light from a glass roof. Spray from the ceiling water pipes caused a visible dampness. The moisture wet the warp ends and reduced their tendency

[5] One could also ask why did we study textile industry. Obviously, there was a greater probability of obtaining necessary permission from Polish authorities since the textile industry was not related to national defense.

to break. The workers hated the moisture; however, dryness also produced complaints because of the frequent breaking of the warp ends. The noise was so great in the shop that one had to raise his voice considerably and speak into the ear of the other. On the whole, it was sweaty, noisy, sticky—still miles away from a pleasant shop atmosphere and environment as imagined, for example, by John Ruskin.

Our Group I worked close to the main shop entrance, facing the office of the shift supervisor. In contrast to American practice, the office was separated from the shop by a solid wall, instead of a glass panel, so that from his office the supervisor could see neither the weavers nor the foremen.

Group II worked right behind Group I, in the same section, which was in the corner of the large shop, where they had easy access to a "smoking room," the antechamber of a shower room used only on Saturdays by foremen and supervisors. Sitting there with them, while they were resting and smoking, gave me a wonderful opportunity for talking and listening.

A work group consisted of twelve persons, who attended a total of fifty-four looms.[6] The looms of the work group were in three rows, eighteen looms to a row, placed very close together, except that the first twelve looms (which were quite large) were divided from the other six smaller looms by a corridor. In each row of eighteen looms were four workers— one person for every four of the larger looms and one person to attend the six small looms across the corridor. There was no spatial division between Group I and Group II. In Figure II persons are identified according to the work group by roman numerals I or II. The arabic figures following the hyphen identify the individuals, with uneven numbers for men and even numbers for women. Thus, II-9 identifies a male worker

---

[6] Some women were not efficient enough to operate six small looms at once. Therefore, in some groups, there were two women attending the six-loom set. This accounts for the fact that some work groups had more than twelve persons. In our case, Work Group I usually had thirteen persons. However, because one worker was absent because of sickness during our observation period, we dealt with twelve persons in Group I.

in Group II; I-0 means that that particular set of looms was not attended most of the time of our inquiry.

Restrooms both for men and for women, were in places quite removed from the shop, out of sight of the section foreman or the foreman. Nevertheless, men mostly rested in the "smoking room," and women once in a while walked out of the building, passing the office of the shift supervisor, to stand in the fresh air for a few moments. The women's dressing room was on another floor of an adjacent building; men changed in the shop, frequently in the smoking room. The foremen had a special dressing room at the other side of the weaving shop.

FIGURE II

LAYOUT OF THE "CORNER" OF THE WEAVING SHOP

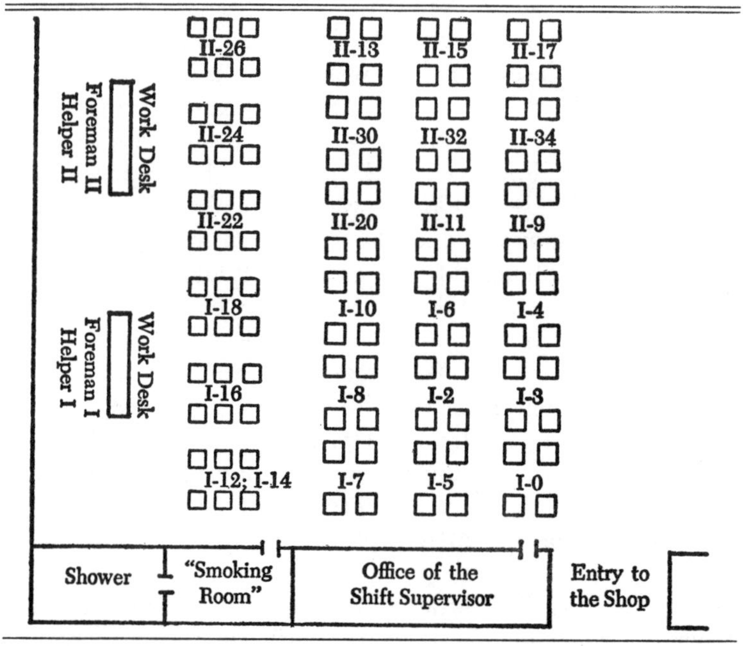

Behind Work Group II was Work Group III, and so on, altogether eight work groups along the wall of the shop. These were supervised by the section foreman. Beside this section was another, set off by a corridor. For some looms there were three-shift schedules; for others, two shifts. Our work groups belonged to the weekly rotating three-shift (swing) schedule. The eight-hour shifts began at 5:30 A.M., and the looms turned without interruption for twenty-four hours a day, five days a week. When we first contacted the men and women, they were on the second shift, that is, the afternoon shift. On Saturday, people worked six-hour shifts, finishing the whole schedule at midnight. At 11:15 A.M. on that day, the humming of looms stopped for fifteen minutes, and the workers cleaned the machines. Otherwise, there was no interruption for the looms all the week long, unless some technological trouble stopped them temporarily. Usually, there was no work on Sundays.

When a worker took over the looms from his predecessor, the selvage of the material on which he was starting to work was stamped by the foreman. Thus, he continued on the same material at the same speed as long as the machine kept running. Because the machines were old, they often broke down. The most frequent cause of stoppage was the breaking of warp ends.

Each work group had its own foreman and helper. The helper was usually an elderly woman with long experience on the job. Her principal duty was to help the weavers to tie broken warp ends. The foreman, on the other hand, was in charge of loom repairs. If several weavers needed help, they had to await their turn. The foreman and the helper were also the most frequent transmitters of news.

Although work continued without interruption for eight hours, weavers left the job at intervals for short periods, but there was no organized rest period during the whole work time.

The foreman had little authority. The section foreman supervised all eight work groups with their eight foremen. When a worker was to be shifted to other looms or wished to leave early or to go to see a doctor, he had to clear with the section foreman. Thus, the foreman's function was principally to help the weavers. If his status was higher, this was due usually to his greater skill and to his greater access to higher echelons of the management.

The section foreman kept records of production and absenteeism. He also advised foremen in technical problems connected with repair, but rarely handled the machines himself. He was no longer doing "dirty" work. Characteristically, workers felt that he belonged to "them," that is, to the management.

At the next level was the shift supervisor, who had under him several section foremen. He rarely left his office. Officially, the workers were supposed to see the shift supervisor if the section foreman did not satisfy them or could not help them; however, in most cases they went directly to the manager of the weaving department.

Of the three levels of supervision in the weaving department (if we exclude the foreman), the middle level, that is, the shift supervisor, was from the viewpoint of the workers relatively least important. However, it is quite likely that this was due not to the position but to the type of person who was occupying that position. Because the manager of the department was popular, workers tended to see him if something could not be handled by the section foreman.

Persons were paid on a piece-rate basis. The pick clock measured the material they produced. How much they got for a unit of their product depended upon the rate given for particular material and the particular method of producing it on particular looms. The norms for financial rewards were determined by the headquarters of the textile industry in co-operation with other governmental agencies and the Central Committee of Labor Unions. The worker himself had no influence upon the establishment of norms. The men and women did not know where, how, and in most cases by whom these decisions were made.

Before the October change, the plan of production was more

significant because larger portions of premiums of the foreman depended largely upon the amount produced by his group. In addition, weavers who were outstanding producers were rewarded by premiums, a measure quite unpopular among the mass of workers. Previously, the guaranteed wage for the foreman had been 900 to 1,100 zlotys, and all his other compensation depended upon the fulfillment of the production plan by his weavers.[7] According to the new policy introduced in January, 1956, the foreman received his guaranteed 1,500-1,800 zlotys, depending upon the number of looms, and 4 percent of the earnings of each of his weavers.

Production quotas were worked out in an interesting manner. The headquarters of the textile industry sent to the factory a rough outline of production goals to be reached. The planner, who worked under the manager of the weaving shop and the supervisor of the accounting office, broke the production figures into more detailed plans, taking into account the particular variations in the shop. Then, in cooperation with the manager of the weaving shop, he worked out weaving quotas, developed for each quarter of the year, based on past plant experience. The detailed plan then went back to the headquarters of the industry for approval.

According to the planner, there were between twenty and thirty different types of material for which the rate of production varied. There were, furthermore, variable rates for three different types of looms and for three different types of operations of the looms. Thus, altogether there were more than two hundred different combinations of possible factors that might account for greater or smaller norms. In practice, however, the variations, happily for the planner, were actually not that many.

Every weaver, therefore, had his quota for each type of material, and he knew by how much he was ahead of or

7 The official rate of exchange was 24 zlotys to one United States dollar. Privately, persons were offering 100 to 125 zlotys for one dollar.

behind the plan. Every quota developed by the planner gave an allowance of 4.5 percent for "downtime" on the machines.

According to the planner, only once had persons complained because of the high norm of the plan. There was, however, one area of disagreement. When there was a cessation of work that was due to causes beyond the weavers' control (for example, failure in the power), the workers were entitled to additional paid allowances computed on the basis of their average production. How much time was allowed for this cause depended upon the estimates by the foreman and the section foreman. This obviously contributed to the raising of status of the foreman, who otherwise did not have much to say about the work.

For overtime, the weaver was paid his average piece rate plus 4.75 zlotys for every overtime hour. This amounted practically to double the normal pay of most workers, according to I-5 and Foreman I.

From the wage was deducted the tax and the fine for spoiled material. The perching unit of the weaving department determined which faults ought to be ascribed to the weaver. Here, of course, was another area of possible disagreement.

Finally, let us mention that the wages were paid on the last and the fifteenth days of the month. The payment in the middle of the month was an advance on the final payment, when the fines and other adjustments were included. The wages (in banknotes) were placed in an envelope.

Each work group had its own representative, a trustee who was supposed to handle labor union affairs. In Group I this was the man identified as I-5, and in the second group it was the woman II-22. The trustees' job was to collect membership dues. According to them, no other duty was assigned to them. As II-22 commented: "Formerly, I was a helper. Then, the section foreman asked me to become trustee. Nobody else wanted it, I took it." "What did you do?" I asked her. "I collected the dues and in return gave the people labor union stamps. I do not do it anymore, because since this spring the dues have been deducted from the paycheck." "Did the other persons agree?" "Oh, yes, they were glad I took it."

In Group I, the trustee was not a helper, but worker I-5. We shall hear about him many more times, because he was the delegate to the workers' council. According to him, he was asked directly by the weaving department labor union secretary to serve as trustee within Group I.

In the group we studied, the section foreman was the only member of the party. But of the management people, the majority were party members, including the shift supervisor, the weaving department manager, the weaving department chairman of the workers' council, and the weaving department labor union secretary. Of those who were classified by workers as "they" (management), the majority belonged to the party. I-5 was approached by the party secretary and asked to join, but he had not so far made up his mind. I-16 was a member prior to our arrival; she reported she had been expelled because of her failure to attend meetings and to pay the membership dues. In the whole weaving department, 12 percent of persons were party members, according to the weaving department party secretary.

The workers' council was the newest and latest organization within the factory. Compared to other organizations, it had aroused great expectations among the workers. I-5 was elected through a free process in which the majority of workers participated rather spontaneously. At the time of our arrival, certain disappointments were appearing. Here before our eyes the problem of participation of the employees in the management was developing like a drama.

Figure III outlines the chain of communication and authority in the four organizations whose task was to lead and serve the workers. It appeared on the basis of reports and observation that contacts occurred most frequently between the persons or groups connected by the solid lines in the figure.

FIGURE III
STRUCTURE OF ORGANIZATIONS

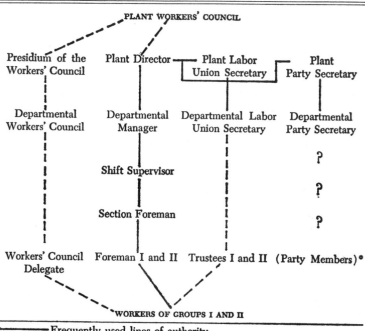

————————Frequently used lines of authority
— — — — —Rarely used lines of authority
        *No party members in Groups I and II

In Figure III the workers' council is above the plant director and plant party secretary, for according to the law, it was supposed to be the highest authority in the factory. Actually, though, the highest power was wielded by the plant director.

Comparing the workers' council with the other three established organizations, it should be noted that a delegate to the workers' council could also become a member of the presidium while remaining in his position of weaver. Within the workers' council, only the position of the chairman was a paid job. The labor union and party departmental secretaries, however, were full-time paid jobs, as were all other positions above them. Thus, the workers' council was more open for vertical movement of persons and ideas than the other two organizations.

Note that the management had two levels of supervision between the manager and the worker. The other organizations occasionally used the management channel to communicate with the workers. For example, the news about some excursion or a workers' council meeting would be transmitted by telephone to the office of the shift supervisor and he would ask the section foreman to tell the workers. The fact that the management personnel handled frequently the news of other organizations contributed to overlapping of the organizations in the mind of the workers, as will be shown in following chapters.

Let us now introduce in a rather brief way some of the more important men and women whom we had the pleasure to observe, to talk with, and to question.

Foreman I was a short, stocky man. He had an outgoing personality, but was easily angered. His willingness to aid the workers was spontaneous. Highly patriotic, he liked to moralize about the Polish nation, speaking sometimes in a pathetic way: "The Polish nation is an enemy to itself. We fight each other. I tell you, sir, other nations stick together but Poles don't. Only when the situation becomes dangerous do they start keeping together. Of course, my dear sir, our nation has always been suffering, under occupation, divided by foreign powers. We have not had enough time to educate the people. But the Polish nation is a good nation; what it needs is this: peace and education."[8]

Foreman I was born in Lodz and started to work in the textile job at the age of sixteen. He had attended school for seven years. He was forty-seven, and he had one daughter in

[8] It should be remembered that to them I was a foreigner, and therefore, the nationality has become more significant. A researcher who was a Pole might not elicit such strong display of nationality identification.

primary school. His major experience in life was his arrest by the Gestapo during World War II because, as he said, he was defending the honor of his wife. He was severely beaten, having some of his teeth knocked out. He gave this as a reason for his present occasional tension and lack of balance. He was quite frank about it: "I know people have complained about me because of my hotness. But I have explained to you how I got my nerves ruined. I am a good specialist, but sometimes nervous." He was very fond of his daughter. When I had an interview with him in the church park, he brought her with him. He said that he would like her to get a better education than he had. Eventually he hoped that she would become a medical doctor.

I-5, the "delegate and trustee," was born in a village some thirty-four years earlier. He was the father of three children. During the war, he participated in the underground movement as a member of the Peasants' Battalions. No one else from the men's group had taken part in any underground movements. Any arrests on their records were for personal reasons—escape from forced labor, black market activities, and the like.

I-5 was handsome and possessed good speaking ability. He was intensely interested in our work and cooperated gladly. Because of his workers' council position, he felt some responsibility for all that happened in the shop. He also had enough personal courage to speak out. While speaking, he gesticulated, eventually becoming heated over some issues. He felt insecure when dealing with persons more educated than himself. We encouraged him to read, bringing him books and later inviting him to the Sociological Institute. Because of his village origin, he tended to compare the situation on farms with that of the factory worker. He was probably the only man who tended to think of contemporary Polish problems on a nationwide scale instead of the limited perspective of the other factory workers. Because his wife also worked, he was able to make ends meet in his family budget. He was not the top producer, however, for he liked to spend time in the "smoking room," debating and asking questions. On the whole, he distinguished himself from the other men. Frustrations which he experienced and complaints he spelled out were exceedingly helpful to us in our efforts to understand the problem of the workers' council.

Foreman II was forty-two years old, married, and the father of two children. Born in Lodz, he had seven years of school attendance and had been working in the factory for twenty-five years, that is, even under the German occupation. He lived in one of the red-brick apartment houses owned by the plant. His wife worked in the factory also. Foreman II was polite, self-controlled, and reticent, and was just the opposite of Foreman I. He kept his group running well, and one could frequently see him just walking around at ease, joking kindly with the women, while Foreman I was perspiring and hastily repairing machines. However, he did not tell us anything about his opinions and feelings. He was also very suspicious about my identity, and up to the last moment he never seemed to trust me completely. He also did not have the strong "Polish nation" identification, observing events predominantly from his personal viewpoint.

The section foreman, whom we have already described as a tall, good-looking man, in his forties, married, and the father of three children, was born in a family in which there was a teacher and had himself graduated from a technical high school; he had also enrolled for a few evening courses at a technical university in Lodz. During World War II he was a prisoner of war in Germany. Because he came to Lodz from the eastern territories, those persons who did not like him maintained that he was "a Russian," but he was Polish, of course, a member of the party. He used slow but articulate speech. At the beginning distrustful of my research, he later cooperated quite willingly. He knew everyone in his section—about a hundred persons—by name; he could tell about their weaknesses as well as their good points. The section foreman appeared always neat and composed, despite the fact that he confided that sometimes the people racked his nerves. If he could, he said, he would like to finish his studies or to start a small enterprise of his own.

The short and rather fat shift supervisor was a cautious man. Workers felt no ill will toward him, but somehow he did not count too heavily in the hierarchy of power. He had been born in Lodz in a family which had one textile worker. Since the age of fifteen he had worked. During the war he was a forced laborer in German mines. Now married, he had two daughters and lived in an apartment consisting of a kitchen and two rooms. (The section foreman reported himself as having only one room.) Before the end of the war, he had been in hiding in Lodz, later joining the Polish Eastern Army. After demobilization, he became a foreman. Because he was efficient, he was sent to the textile school. Later he was chief engineer in a textile plant in Polish western territories. He reported that "because climatic conditions were not suitable to him there," he had returned to Lodz. He was a party member, and generally he was described as a man who held his position because of that membership. However, he was also a man for whom the new political-social system provided an opportunity for education and social advancement. He stressed that before 1939 it would not have been possible for a foreman to go to school and to advance socially. According to him, workers did not sufficiently appreciate certain opportunities and improvements brought about by the party. However, when there was a conflict between the management and workers, as we shall show in connection with the issue of the so-called 13th wage, he wavered between both sides.

The manager of the weaving shop was a calmly speaking, married man in his thirties, father of two children. He was born in a village, the son of an artisan. During the war he had worked in Lodz as a weaver. After the war, he went to the textile school. Later, he was the labor union secretary of the weaving department, a job he had held for the past two years. Even more important was his membership in the Directorate of the Textile Industry for all Poland with headquarters in Lodz, and in the assembly of the Lodz committee of the party.

The manager was intelligent; he listened carefully, and then he suggested what would be best for our research. As a matter of fact, he was liked because of his listening ability and "treatment of people." When he could not comply with a demand, he always gave a reason, and this was appreciated by the workers. The fact that he did not shout at people was also appreciated. The manager was the only man who opposed the plant director on several issues. Since the plant director was unpopular with weavers, this contributed the more to the manager's popularity. When differences of opinion between both men finally resulted in the transfer of the manager to another factory, employees said that the plant director had won over the manager. They felt that this was a loss for them and that times might again get worse. This competition for power between the plant director and the weaving department manager was of great interest to us, and we shall describe it in greater detail later.

The chairman of the workers' council weaving department held an elective position. His paid job was oversight of thirty-eight clerical workers, a staff position immediately under the manager; his occupational title was "supervisor of the accounting office and bureau."

He was born into a family of a textile worker in Lodz; his mother and wife were both working in other factories. In his forties, he had spent the war at forced labor in Germany, remembering when hunger drove him to eat raw wheat in the fields. Returning to Lodz, he went to school, receiving training in technical administration. His son wished to become a surgeon. He was gentle and kind, but lacking the drive of an executive. He was rather skeptical on different issues.

The weaving department's party secretary was a former rank-and-filer who had risen to his present position through attendance at a special party training school in Torun. He was married, and his wife was working in another factory. He was in his late thirties, outgoing and friendly, sensing human problems. He spoke slowly, smiling broadly. However, his formal education was rather limited, and he frequently with embarrassment dropped an argument when it reached higher levels of abstraction. He said that he would like to go back into production where he would earn more, but his party secretaryship was a paid full-time job, and he felt dutybound to perform it. Speaking about people in the shop, he said: "They want many rights and little duties." Challenging me on another occasion, he maintained, "We have many persons who are too old and inefficient in the shop. In a capitalist country they would be dropped. But we give them employment." From my observation I would concur with him that many persons in the shop did not work efficiently. However, there was a shortage of labor. Were not these older and inefficient workers employed primarily because of this shortage?

The party secretary described his role in the weaving department in the following words: "When something goes wrong, the party acts. Members of the party are supposed to have a higher responsibility. Personally, I don't like to interfere. We do not like to dictate. If somebody comes with a complaint, I send him to the labor union."

The weaving department's labor union secretary was a woman in her thirties; she had been formerly a weaver herself. She was married, but we did not obtain any further information concerning her. She was a party member, not very popular with workers, although the lack of popularity might have been due to her position rather than to her personality. The labor union role she described as follows: "We look out for the welfare of workers, taking care of their housing problem, tram and railway reduction tickets, organization of excursions, and so forth."

The plant director was a short, stocky, married man in his late forties; he had been formerly a weaver. He belonged to the category of workers who had distinguished themselves by initiative and political awareness and had made careers in the post-World War II Poland. However, reportedly he never had been a foreman himself. When dealing with people, he smiled benevolently. When speaking at meetings, he raised his voice. The workers disliked him because of his "shouting." On the whole, he was unpopular. We shall hear how during the October crisis several persons had sought to remove him from his office. Presumably, the party had saved him. However, he was not a cowardly man. I saw him surrounded by a whole crowd of angry workers and yet seeking to win them to his viewpoint.

The chairman of the workers' council presidium, who was also chairman of the workers' council, went on a health vacation at the beginning of our stay in the factory; later he held a supervising job. His differentiation between the role of the labor union and that of the workers' council deserves to be recorded: "The labor union is supposed to deal with the welfare of workers, while the workers' council has to be concerned with the production." He also provided us with information on the beginning of the workers' council: "We talked about that in October; we listened to Gomułka's speech on the radio. He said that we should create workers' councils. Thus we discussed it and formed a committee. The committee proposed candidates." The chairman stressed that the candidates were freely elected, that the workers themselves decided about persons.

As for other top-level personnel, we obtained no personal information concerning the plant party secretary and the plant labor union secretary. We did not have many chances to see them. The men and women of our group met the party and the labor union through the weaving department's secretaries.

Of the twenty-four weavers in our two groups (excluding the foremen and helpers), eleven had attended primary school in a village, seven in a small Polish town, and four in a city; two had had no schooling at all. The average number of years of schooling for persons in the group was four and a half. The origin of eighteen in small towns or villages illustrates the well-known movement of rural population to industrial centers in post-World War II Poland.[9] We discovered that not one of the persons studied had a father whose occupation was non-manual. Of the two women who had had no schooling at all, one of them (I-12) had learned to read by her own effort, and the other (I-18), an unmarried deaf-mute in her late twenties, was taking special instruction in reading and writing at the time of our research. Seven of the twenty-four had been drafted into forced labor in Germany during World War II.

The average age of the workers in the group was thirty-one years, and the average length of time as weavers was five years and two months. The workers had been employed in their present groups an average of a year and seven months. The average monthly income for the twenty-four persons, computed from their net monthly income from January, 1957, through June, 1957, was 1,122 zlotys.[10]

The foremen, of course, had had considerably longer job experience than the weavers. Foreman I had been employed in the plant since 1935, and Foreman II since 1933. The elderly female helpers had had the longest work experience of all—Helper I had worked for fifty-six years and Helper II for thirty-three years, all of this time in this same plant.[11] The foremen had an average income of 2,300 zlotys, and the incomes of the helpers were about the same as those of the weavers.

[9] In 1946, there were 16.1 million people living in the country and 7.5 million in cities. In 1956, the corresponding figures were 15.5 million in the country and 12.6 in cities. Główny Urząd Statystyczny, *Rocznik Statystyczny, 1957,* xxix.

[10] As usually both husband and wife worked, the family income amounted to over 2,000 zlotys in most cases. For example, I-2, a mother of two infant children, worked on different shifts from her husband. Thus, they managed to be also able to stay with the children alternately. Of course, the one working the night shift could not sleep much during the day and so for that week was extremely tired. Although the factory provided a nursery for children of working mothers, I-2 disliked to put her two children there.

[11] Helper I was seventy-two years old and still kept working, eager to increase her retirement benefits. She was a small, tiny woman, with silver hair, a "babushka" who remembered "the good times" of her young years under the Czar. She was not taken seriously by other members of the group.

Of the twenty-four persons, eighteen were married, one was a widow, and one was divorced. But the divorced woman was determined to win back her husband. The number of children averaged 1.5 per family. Sixteen of the workers lived in one-room apartments; only seven had two-room apartments. Fifteen of the twenty-four had radios. The average expenditure on living essentials was: rent, 43 zlotys; food, 1,070 zlotys; and clothing, 278 zlotys.[12]

I discovered that one-room apartments were living space and kitchen combined, and sometimes single persons shared the room with a family. One weaver (II-30) reported that she was sharing a room with a couple with two small children; this was a private arrangement, and her rent was 100 zlotys, more than the official rate (rents had been frozen by government decree). These appalling housing conditions make it clear why for almost all persons in the group the problem of more living space was uppermost in their minds.

With such a high proportion of their income going for food, the members of the group would, according to most standards, be on a subsistence level. It should be pointed out, however, that Polish workers in 1957 were afforded free medical treatment, a paid vacation, and a retirement pension. In addition, there were free educational opportunities, and various sorts of tram and railway tickets and personal services, such as haircuts, were kept at low rates. On the other hand, the prices for commodities were relatively high.[13]

[12] The average expenditure figures show a greater sum spent than the sum earned, as reported above. This is due to mistakes of three respondents who included in their data expenditures from the income earned by their spouses. This inexactness in our data does not, however, affect their meaning, because we are here interested in the relative spending of the income, whether earned by one or two persons. It must also be reported that from the twenty-four respondents, only sixteen persons answered the expenditure question; furthermore, the sixteen persons did not always supply data for all three expenditure categories.

[13] According to a married respondent from work group III, one kilogram of meat cost 30 zlotys; of sugar, 13 zlotys; of bread, 3.7 zlotys. A man's jacket and pants cost 1,300 zlotys—more than the average monthly income of a weaver —and a pair of shoes, 400 zlotys.

The cinema was the most popular pastime among our group, and church attendance on Sunday morning a prevailing practice. Only three of the group never attended a motion picture, four attended movies only once each quarter, but seventeen persons of the twenty-four attended movies more than once a month. Nineteen attended church every Sunday, three persons on holy days, one person did not answer, and one said he never attended religious services. These leisure attributes are typical for what we know of Polish workers, especially those recruited from rural districts. All the respondents were Roman Catholic.

The last item of our inquiry on the workers' subculture was the issue of their aspirations.[14] Note in Figure IV that the

FIGURE IV
FUTURE ASPIRATIONS

1.	Which would you prefer:	(a) A new apartment	15	persons
		(b) A raise in wages	9	"
2.	Which would you prefer:	(a) To keep the same job	13	"
		(b) To change your job	11	"
3.	If you want to change your job, would you like:			
		(a) To become a foreman	4	"
		(b) To get a white-collar job	1	"
		(c) Not to work whatsoever	3	"
4.	Do you want your child to become:			
		(a) A surgeon	4	"
		(b) A graduate engineer	5	"
		(c) A teacher	2	"
		(d) A skilled artisan	2	"
		(e) A film star	1	"
5.	If you were suddently to win some thousands of zlotys,			
	would you:	(a) Construct a family house	2	"
		(b) Buy a new apartment	5	"
		(c) Buy new clothing	11	"
		(d) Buy new furniture	1	"
		(e) Buy a motorcycle	2	"
		(f) Buy fuel and food	1	"
		(g) Buy drugs	1	"
		(h) Save it for your children	1	"

14 In the "future aspirations" questionnaire, the questions were not necessarily exclusive. Therefore, the sum of answers might be more or less than twenty-four.

respondents themselves did not aspire to move to white-collar jobs! On the other hand, vertical mobility aspirations appeared clearly as far as children were concerned. It is noticeable that no one listed the priesthood as desirable for his children. One should remember that over half of our respondents were reared in the country, where the traditional occupation to be aspired to was the position of the Catholic priest. Only two respondents wanted their children to remain on the worker's level, though a skilled worker in a trade. Otherwise, all parents expressed a desire for a professional job for their children. In terms of social mobility aspirations, this would make for a tremendous vertical mobility in Poland.

The desire to spend windfall money on immediate material needs, especially on dress, indicates the standard of living of the people studied. It should be pointed out that the first item in the questionnaire referred to the choice either of raise in pay or of a new apartment. As the reader will remember, the majority of people preferred a new apartment. Thus, at the bottom of the same questionnaire, the respondents naturally did not list the apartment as frequently as they did when first asked.

Within the windfall money question, they were asked also whether they would like to spend the money on travel, or the buying of books, and so forth. There was a clear concern with tangible needs, while interaction and cultural needs were listed practically not at all.

3

Past Events in the Factory

AS A RESULT of the dramatic and bloody Poznan riots in June, 1956, Poland was internally shaken. The de-Stalinization process was accelerated by great economic dissatisfaction of workers on the one hand and by moral indignation of intellectual groups on the other, rejecting for the first time in public the misrepresentation of the Stalin period. Added to this was the traditional anti-Russian attitude of the Polish masses and the strong influence of the Catholic church. The group of Natolin being defeated, Gomułka was swept into power. "He suffered under Stalin; he was in prison"—so said the man in the street. Lodz was convulsed by great excitement. The news about the Hungarian revolution aroused emotions. Workers coming to the town from nearby villages reported seeing tanks posted around the city.

It is interesting to note that Lodz, the second major industrial center, kept relatively quiet when the Poznan riots broke out in June, 1956. But during the "October Revolution," the city and its predominantly industrial workers experienced a high wave of hope for social and political change. "The enthusiasm was tremendous," said one of the clerical workers. According to the weaving department party secretary, "The party lost its contact with the masses in Poznan and in Hungary; and in order to prevent loss of contact, it is necessary to do what the masses wish."

Now my informants could not remember accurately the flow of events in the dramatic October days. From nineteen persons who themselves participated in the October events within the factory, I have pinned together the story.

The newspapers and the radio alerted people to the great events. In the shop, the first unusual thing which happened was a mass meeting of the employees.

I-6: "Gomułka got into government, and the people were released from prison. At once a mass meeting was organized. The foreman told us that we must go to the meeting. All of us went. People shouted: 'Out with everything.' Long live Gomułka!' Somebody from the town spoke. The plant party secretary said that Gomułka was in power. We shouted, 'Long live!' This was a short meeting. Then we got back to work."

Foreman I: "First there was a short, ten-minute meeting, organized by the management and the party. They told us about the new government headed by Gomułka. The plant party secretary told us that this was a manifestation meeting."

Although the meeting was organized by the management and the party and the plant party secretary spoke, according to all evidence, on this occasion, the party genuinely expressed the general will.

For two weeks several departmental or party-member meetings were held. The last meeting to which respondents referred was a belligerent preelection meeting, attended by the two Lodz party candidates for the Polish parliament.

Meanwhile, feelings against the plant director were expressed by several persons. Respondents reported that the plant was slow in removing the red star, a symbol of foreign and Stalinist domination. The stocky plant director was held responsible for this fact.

The young section foreman: "A delegation of four students contacted the management and the party to get permission to organize a discussion meeting with workers.[1] This meeting was held at the time of exchange of work shifts, and attendance was voluntary. The plant party secretary chaired the meeting, attended also by the newsmen from the town. On this occasion, even those who had never before raised their voices spoke out. Complaints and mistakes were disclosed. The students attempted to explain what was going on. They asked the laborers to keep themselves restrained and reasonable. The manager of the weaving shop also spoke, trying to calm the people down, advising them not to start street riots."

I-6: "At this meeting, people were against the director. They accused him of depriving them of premiums, of wasting money. And he asked people for forgiveness, blaming his mistakes on the lack of a good education. You know, sir, the meeting had to be stopped. He did not want to have the star removed; therefore, students went to his apartment and took him with them."

Foreman I: "People wanted to have a meeting. They asked us when it would be. An unsigned letter was sent accusing the plant director. Therefore, a committee was formed to investigate his activities. One weaver spoke; it was a noisy meeting. People asked that the plant director and all around him be removed; there was quite a lot of turmoil."

The chairman of the workers' council presidium: "The plant director is efficient. Before his arrival, the factory was not a top producer. Now, we have succeeded in winning the banner as the best factory."

The above statements show that the factory went through a wave of dissatisfaction that did not really fully develop into the removal of the plant director. It was as though a lid had been removed. The pressure of old resentments and complaints blew up. I saw on another occasion that workers

---

[1] Note the leadership role of university students. Within the Polish culture, students are generally expected to act in political and other emergencies, offering leadership and self-sacrifice.

enjoyed throwing their complaints upon the plant director. Why did not popular dissatisfaction remove him as had happened in other plants in which the labor groups eventually dragged out the management from the factory in a direct way? The following explanation was offered by a woman from another group: "The plant director was backed by the party. The party asked the workers to postpone their decision, that the issue would be discussed later."

Another plausible explanation could be a lack of any organization among the workers themselves. There was only one report that a group of foremen sought to become organized in order to effect the removal of the plant director. Otherwise, the workers did not develop any organizational action of their own. Interestingly, we shall later report events in which workers undertook themselves some steps as a group. This was done, however, within the department only. A factory-wide organization was lacking. Moreover, both the students and the manager of the weaving shop, who enjoyed some prestige among the workers, were asking for calm, playing down the excitement.

While the public meeting with the plant director and students was organized, several smaller meetings on the departmental level were reported, and, of course, party meetings. Foreman I: "There was a party meeting. The members were asked to quiet down the people, to tell them that everything would be settled later."

Meanwhile, an investigating committee inquired into the plant director's operations. The section foreman was a member of the committee. According to the supervisor of technical control, the committee found no evidence to support charges that the plant director used factory material and factory people to repair his private apartment.

The last meeting, vividly remembered and reported by many participants, was a preelection meeting, addressed by the party's Lodz candidate for the Polish parliament.

Foreman I: "The preelection meeting was organized by the

management. It was held during worktime and was supposed to be attended only by those shifts which were not at work. However, some people left their looms and attended the meeting. There were leaflets distributed after the meeting, asking people not to vote for them."

The supervisor of technical control: "People shouted, 'We don't want you!' But she answered, 'If you don't want me, don't vote for me.'"

II-32: "There were great complaints. People complained that nobody had paid attention to their complaints before. She said, 'Though you don't like me, the party wants me.' And this brought a great outburst in the hall. He asked, 'Why don't you want me?' 'Because you did not do anything for us,' people shouted, 'and the radio said that Gomułka will support workers, that everything should be given to workers.'"

II-17: "She said that everything would improve. But I told myself, and the people also said, 'Well, let's wait and see. I want to see it with my own eyes.'"

Obviously, the preelection meeting was stormy, being prematurely concluded. Nevertheless, within the contemporary Polish election system, it was possible for both candidates to be elected. As a matter of fact, people voiced their dissatisfaction and anger without translating these attitudes into any collective action. It is quite possible that many of those who openly opposed both candidates at the meeting had later at the polls left their names standing on the unified list of the Polish election of January, 1957.

One feature stands out: there was a great upsurge which within a few days disappeared. Furthermore, the party quickly anticipated and channeled popular dissatisfaction into verbal protests at meetings. Note that all meetings were organized by the management and the party. Thus, a somehow paradoxical situation occurs: the group that is the target of criticism and dislike organizes the opportunity for its own exposure. Even the protest is organized by the party and within the party. Characteristically, the labor union had nothing to say during the whole dramatic October.[2] The organization that was supposed formally to express and defend the interests of workers was not once referred to by any of my informants.

What was the result of October events within the factory in addition to the relief-providing public expression of complaints and dislikes? Within the top management, no change had been reported. However, in lower echelons, some personnel changes occurred.

The supervisor of technical control: "One shift supervisor was demoted to section foreman."

The weaving shop manager: "There were several section meetings. People asked that some foremen be removed. We complied in two cases; one case we did not accept."

As stated above, the general excitement soon receded and life returned to its normal course.

I-4: "At the beginning there was a great applause for Gomułka, a great rise in hopes. Now they don't applaud any more. He released persons from prison, he raised salaries, and they still want more."

However, such a statement was rather unusual. On the whole, high expectation naturally resulted in disappointment. Yet there was one gain.

The shift supervisor: "There has been greater personal freedom. Before, if something went wrong, anybody could have been called responsible and classified as 'a people's enemy.' "

From my own observations, the increase of free expression was a fact, at least during the summer of 1957.

Another innovation was the introduction of the workers' council. Reviewing the rules of the workers' council of the plant, we find that the membership was set at 113 members, from whom a presidium of 15 persons was elected.[3] Happily,

---

[2] Renata Tulli, "Pozytyvny czy Negatywny," *Nowa Kultura*, VIII (June 30, 1957), 2. The author has found that neither the labor union nor the youth organization played any significant role in October, 1956.

[3] See the translation of the workers' council bylaws in the Appendix.

I-5 was elected even to the presidium, providing us with an insight into the operations of that body.

The workers' council bylaws said that time spent on council activities should be paid at the rate equaling an average of the members' normal earnings. The rank and file could attend meetings, provided that this would not interfere with their duties. The role of the presidium was defined as the executive body of the council. But it was also accepted that the management of the whole plant was one man's job and one man's responsibility. If necessary, the plant director was entitled to undertake certain steps without first consulting the presidium. However, he was requested to give an account of such decisions at the next meeting of the presidium. Furthermore, it was stated that in case the director might disagree with the decision of the workers' council because he found it incompatible with legal rules, economically disadvantageous, or contradictory to the plan, he could refuse to accept the decision. If no agreement could be reached, both or either of the parties could appeal to the Ministry of Light Industry. The clause providing for a veto of the workers' council decision by the director therefore actually left control of the enterprise with the governmental headquarters in Warsaw.

Now to turn back to our account of events. Several respondents told us something about the beginning of the workers' council. Again, we will present just a few of the sixteen statements we obtained.

The supervisor of technical control: "The preparatory committee for the workers' council was organized by the party, assisted by the management. The manager of the weaving shop was the chairman. They set down the rules under which the election should be carried out. Then, within all departments, election committees were formed. They organized polls on the departmental basis. All procedures were recorded."

Note that again the management and the party had initiated the action for the workers' council. Let us see how the election process occurred within our two groups:

II-17: "The section foreman told us that there would be a preparatory election meeting after work. At the meeting there was the office supervisor. He had already been chosen as chairman of the committee. We discussed the organization of election, the selection of candidates. Every section should offer its own candidates. One said that the workers' council would be the organ of the workers' rule over the management. This is what one said at the beginning."

The young section foreman: "The meeting was organized in the small hall. Every section had its own meeting. The office supervisor was the chairman. The meeting was announced over the radio; one read about the workers' council in the newspapers. The office supervisor spoke spontaneously. There wasn't anything decided in advance concerning the candidates. Directly from the floor, people suggested persons who should be their candidates. There were two candidates nominated at the meeting, Foreman I and I-5. Later, the office supervisor prepared small cards on which were written the names of the candidates. People dropped the name of their chosen candidate in a box. In the dressing room of the foreman, several persons counted the votes and then immediately announced the results."

The election was held during the work period on another day. Let us now turn once more to II-17: "Another day, they came with a box, asking us to cross out two names. There were three candidates, Foreman I, I-5, and a woman candidate from another work group. We had already talked before among ourselves in the smoking room about whom to vote for. Of course, we agitated for I-5. It was said that he was one of us and that he would always stand up for our rights, and we had to choose a worker; a foreman is not a worker any more. Concerning the woman candidate, we felt that a woman does not have enough time for meetings, she must take care of children. The section foreman was also in favor of I-5. Thus, they went along our looms and we dropped in the name. II-34 carried the box. They counted it and told us the result right away."

I-10: "On Saturday afternoon, after the work period, they asked us to attend a meeting. They told us that the delegate whom we would elect would inform us about everything. Then, another day, they went around with the pollbox, asking us to cross out the names of those whom we did not want to choose. I voted for I-5 because he is young. Foreman I could probably forget about workers."

All evidence had shown that this election was fair, serving really as a tool for expressing the majority opinion. We shall see later that dissatisfaction was expressed in connection with the labor union elections.[5]

For what reasons did the people decide to vote for I-5?

A two-shift foreman: "I-5 was elected because he was so hard. He is softer now; perhaps this is because he would like to get an apartment from the management."

I-3: "He was elected because he was a worker himself. I would not vote for a foreman."

II-32: "I-5 is presumably an honest man; he speaks directly and courageously. I heard him at the preparatory meeting. He said that during the night shift there must be steam. It might be that other things have not changed, but we got the steam."

II-24: "I voted for I-5 because he is good. Foreman II told me that he is good."

II-20: "I voted for I-5 because I heard him speaking. He understood the matter. Nobody told me for whom to vote."[4]

II-34: "I voted for him because he spoke well."

Four of these respondents were women. Note that most of them referred to the fact that I-5 spoke well at the meeting. Obviously, the ability to express himself fluently and to formulate and present feelings of the group partially accounts for the leadership role.[5] Also, the differentiation between the worker and the foreman is conspicuous. Many other sections

[4] This respondent could not remember the name of I-5. As a matter of fact, in our quotations we have substituted the identification even if the respondent could not identify a person by name.

[5] In the Polish situation the courage to speak out was particularly significant, since workers were sometimes fearful of those in authority.

elected foremen, however. Probably in our case, Foreman I had certain personality characteristics that prevented his being completely accepted.

It is also of interest to hear the statements of those who were defeated. The woman candidate: "My group selected me. I did not want it, but the group wanted me. I was the only woman candidate. Anyhow, I have not yet heard about one single item that the workers' council would have managed to improve."

Foreman I said that he was happy not to have been elected because his nerves would not have permitted him to go through all the troubles connected with the delegation. He also said that he could not control himself and would tell the management what kind of bad fellows they were. I-5, according to him, was good. He spoke well and had good judgment.

I-5 had attended several meetings of the presidium before our arrival. At one of these, there occurred a dramatic event that deserves to be reported here. On the issue of dividing premiums equally or unequally, the plant director and the manager of the weaving shop disagreed, as on other matters. But underlying this disagreement was a struggle as to who was supposed to be the chairman of the presidium of the workers' council. There were two candidates: the present chairman of the presidium, and the manager of the weaving shop. Let us listen on this particular question to the office supervisor: "The manager was the candidate for the chairmanship of the presidium. He got seven votes, while the present chairman of the presidium got eight votes. You see, the present chairman is according to the plant director's liking; he will not oppose him." Thus, the vote was a victory for the plant director.

The issue of the annual premiums was reported by the young section foreman as follows: "In previous years, premiums from the fund were divided unequally, according to whether the person fulfilled the plan or not. This was unpopular among the people. They anticipated that the same would happen this year. Thus, they asked I-5 not to permit it when the matter came up in the presidium."

I-5 himself said: "Well, I am a little at a loss with the workers' council. It started to function in the spring. For example, on the issue of premiums they wanted the premiums to be divided unequally, that some might get 150 zlotys, some 75, and some nothing if they cut work frequently. But I voted against this suggestion, and the manager voted with me. The manager said that he was willing to give up his own premium amounting to 500 zlotys in order not to have trouble with people. Then, I talked it over with people in the shop. They were fiercely in favor of the equal distribution. We then did it so that the sum given to the weaving department was in our department divided equally. Everybody got 75 zlotys. You know, there is one more point about the unequal distribution. Nobody really knows how much the total sum available for premiums is. Therefore, they can keep the money for themselves; they can cheat us."

The position taken by I-5 was received favorably by his electorate. II-17 said: "Oh, we knew about it. We talked to I-5 and told him what to do." The young section foreman: "I-5 did not call a meeting; he told just a few persons what he had done. But the word spread around quickly."

Since the weaving department had its own policy on the premium, it also had its own departmental meeting of the workers' council, at which time the equal distribution was unanimously accepted. The unanimous vote was reported by the young section foreman.

In this connection, collective action by workers themselves was reported by the manager: "Well, the people were always dissatisfied with the previous distribution of the fund premiums, because only the best workers used to be rewarded. Of course, not everybody has the same capacity to produce, but they wanted to have it equal. Only a few like to preserve the unequal system. Some twenty or thirty persons came to see me, asking me to distribute it equally. They came here, to my office."

It should be noted that this was an action that was not organized by any of the existing organizations. As the workers' council was not yet fully introduced, the people acted on their own, disregarding the party and the line of the managerial command. The labor union organization was not even mentioned by anybody in this connection.

The workers' council had been functioning for just a few months before our arrival. We shall report on its further actions on the basis of the direct observation and participation in the flow of events. Judging on the basis of the above statements, we would say that the feeling about the workers' council was uncrystallized. Disappointments arose, but hopeful expectations were still present. A more definite attitude was expressed in regard to another workers' organization—the labor union.

The labor union was the organization most complained of by workers. It was also an organization that did not play any part in the dramatic events during October, 1956, and following months. Evaluating the role of the labor union organization, Gomułka said that the acceptance of the workers' councils "was largely influenced by criticism directed at labor unions, which also showed signs of sickness."[6] On the whole, there was a demand for redefinition of the role of the labor union.

"In August, 1956, at the plenary section of the Central Council of the Labor Unions, as a result of the wide discussion in the press and at meetings, the proper task of labor unions was formulated as follows: the protection of the vital interests of the laboring people, the struggle for real participation of the workers in the management of the factory and in the political government of the country, the struggle for better wages, housing and living conditions, the organization of the workers' cultural life, and so forth."[7] The reader will remember that the Central Committee of Labor Unions sought to revive shop committees shortly before October, 1956. However, the labor union was discredited. It is probably not quite feasible for

[6] Władysław Gomułka, *Węzłowe Problemy Polityki Partii-Uchwały* (Warsaw, Książka i Wiedza, 1957), 18.

[7] Jan Szczepański, "Poland," in Arnold M. Rose (ed.), *The Institutions of Advanced Societies* (Minneapolis, University of Minnesota Press, 1958), 258.

one organization to achieve two goals, that is, to promote the production and, simultaneously, to protect workers' interests. Probably, both roles cannot be integrated very well within the same body. From this ambivalence resulted the unpopularity of the labor union among workers, and finally its lack of effectiveness.

Our respondents expressed many criticisms and dissatisfactions with the labor union services. We have, however, only two reports on actual events occurring before our arrival. The first is the election of the labor union secretary, and the second, the introduction of the new way of collecting the union membership dues. The following is based upon statements of nineteen respondents.

Concerning the election of the weaving department's labor union secretary, II-13 remarked: "I do not belong to the labor union. It does not give members anything. The present secretary, the woman, has already been sitting there for several years; and she was just supposed to be there one year. Obviously, the party keeps her there."

The supervisor of technical control: "The secretary is elected for two years. Nobody wants to get the position. It is possible that the election was carried out correctly."

It is recalled that each work group had its own union trustee and that this position was not desired by any person. The election of the trustee that took place more than two years ago was described by II-24 in the following manner: "It was at night. The section foreman invited us to his little office in the front. We talked for ten minutes, and I was selected to be the trustee." Other persons could not even remember the election.

The trustees were delegates to the meeting of the labor union at which the full-time secretaries were formally elected. This election took place in January, 1957 (after the October events) and yet there were complaints.

II-24: "I-5 and I were the delegates. We were told that we should again elect the woman secretary. All other secretaries were again reelected. The election was carried out in a speedy way; I could not even think it over. They suggested we should vote for these people, and we did."[8]

I-5, when asked about the way the secretary was elected, also complained that the procedure was too quick even for him. The chairman had proposed the candidates and immediately asked for the vote. Thus, all agreed.

Somehow, the election followed the habitual pattern set before October, 1956. Elections under Stalin were not choices from among different candidates as is customary in the West, but a sort of manifestation of loyalty to the leadership. In this kind of "election," voters naturally lose interest and follow the common pattern of unanimous approval. It is possible that the workers were not much concerned about the labor union even after October, 1956, and, therefore, no one cared much about the election of the labor union officials. This attitude was properly expressed by II-17: "I did not know that there was an election of the secretary. Somebody later said that she was elected and that we should again pay our dues. This does not interest me; we have been paying the dues for so many years."

The possible lack of interest could also be derived from the following statement of the woman candidate: "I have been elected the trustee for our work group. Some people were in favor of the secretary; some were against her. In the corner, there was a group of three persons who had prepared the list of candidates. We voted, and we could strike out the name we did not like. Those who were elected had a meeting later which I did not attend. At that meeting they selected the labor union committee for the whole factory."

There is obviously some contradiction between the statements of I-5 and II-22 and of the woman candidate. According to the woman candidate, the election would have provided for the elimination of the incumbent secretary, were she that unpopular. People sometimes did not use the opportunities offered them by the legal setup, either because they assumed

[8] It was usual to find seated in front of the meeting hall, behind a desk, the plant labor union secretary, the plant director, and the plant party secretary. "They" most likely referred to these three men.

that this would not help or because a certain inertia developed over the years under the tight Stalinist control.

In the issue of the dues collection, we move to a relatively recent event. In the spring of 1957, around Easter, the labor union and the management decided to collect the dues by direct deduction from the paycheck. Workers were supposed to express their approval or disapproval.

I-2: "Earlier, the helper collected the dues from everybody personally. I-5 asked us to come to the section foreman's office and told us that the dues would be one percent, deductible directly from our paycheck. Nobody wanted to accept it. People suggested one-half percent."

II-22: "After work the section foreman called us to his office. We were told about the one percent to be taken off from the paycheck instead of the former selling of membership stamps. Most of the workers disagreed. We wanted to pay only one-half percent."

I-10: "This was sometime in April or May. I-5 walked along the looms and asked us whether we would like to pay the labor union dues. We said that we would pay no more than one-half percent. He probably told the secretary, because later he said that they were agreed. You see, I do pay. If my mother should die, the labor union would help me pay the funeral expenses."

Even if the labor union was conceived as nothing more than a social agency to help in individual problems, we were interested in the labor union role. The most frequent comments were:

I-4: "The secretary stamped my application once for an apartment. The year before last, I spent my vacation in the mountains in a labor union hotel."

I-7: "I have never gone to the labor union with any problem. It is useless. They would not help anyhow."

I-3: "I have gotten only one thing from the labor union, a coupon for potatoes. Otherwise, nothing; they do not give anything."

II-17: "No, you don't get anything from the union. Take,

for example, my child. It is in a summer camp, but this is paid by the management, not by the union. Take the Bicycle Tours of Peace. They told me that there were no tickets available any more. But I-5 found out that even two days later some management people still obtained tickets."

II-34: "Last year, I went to Gdansk [a seaport] for two days. This trip was organized by the union."

II-14: "I have never been in the labor union. Do you think they could help me find an apartment?"

Because respondents were inconsistent concerning their experiences with the labor union, a short schedule was administered. As we had meanwhile discovered that the management also had its own social department, we were interested in finding out to what extent both the labor union and the social department were contacted by members of our group during the past two years. Persons reported no more than one contact for each category of application.

FIGURE V

APPLICATIONS TO LABOR UNION AND SOCIAL
DEPARTMENT DURING TWO YEARS

Labor Union			Social Department		
Kind of Application	No.	Refusals	Kind of Application	No.	Refusals
Loan	6	1	Plant's summer camp		
Coupon for rationed			for child	5	1
wood	3	0	Job change within plant	1	0
New apartment	2	2	Paid vacation	1	0
Stamp on application			Free pass to enter		
to another agency	2	0	and leave plant at		
Participation in			unoffical time	1	0
group excursion	2	0	Stamp on application	1	0
Ticket for sports events..	2	2			
Potato coupon	1	0			
Payment of funeral					
expenses	1	1			
Equipment for infant	1	1			
Total	20	7	Total	9	1

The data in Figure V show clearly the social welfare function of the labor union and also its functional overlapping with the social department of the factory. On the whole, workers did not look at the labor union in the traditional way but rather as a social welfare agency, which, according to the majority's opinion, was not fulfilling its job satisfactorily.

Thus, we meet again the question of whether the workers had any organization they considered as their own. Were the people then able to act as a group?

An incident connected with cheating reveals one set of conditions under which the persons studied were able to develop a unified group behavior. The central figure of the story was I-0, a young man in his late twenties whom we never met because of his illness.

In the affair of the tampered pick clock, let us first listen to I-5: "You know, a few years ago, everybody here was cheating, manipulating the pick clock. Today, this is rather an exception. If I should see somebody doing it again, I would advise him to stop. I would tell him that this would bring him nothing but trouble."

The cheating of I-0 was reported by II-11, while I-5 and I-3 were listening. "You know, when I-0 came back from the army, he was a little queer. He said that no woman was interested in him. He stopped talking. Once in a while he joined us here in smoking, but he did not talk. And one day the section foreman caught him manipulating the pick clock. The police took a record of him and he was supposed to be brought to court. The next day he again came to work, but they would not let him work. Later, the section foreman said if we would guarantee him, they would give him another chance. This we did."

I-5: "I guaranteed him. I told them, 'I did the same some time before. And you gave me a pardon, and you see, I am a good man now. Had I not been given pardon at that time, this would have spoiled my life. I would have joined burglars; I would have gone astray.'"

According to the manager, the flow of events was as follows: "This was before October. And this would mean that I-0 could have been jailed because of stealing. They came here to my office and said that they would see that he did not do it again provided we drop the matter. I told them that all of them would be responsible for him, and they agreed."

The cheating of I-0 promoted an unusual solidarity and collective action on the part of the group. We shall later have another opportunity to see that in an emergency, Work Groups I and II were capable of acting collectively.

4

Eight Weeks in the Factory

BETWEEN JULY 11 and September 3, 1957, there occurred in the factory thirteen events which appeared to have particular significance. These events will be reported more or less in chronological order, with the interpretation proper reserved for the next chapter. The following events will be described:

1. The weaving department production conference on July 17;

2. The production conference of foremen on July 27;

3. The national holiday on July 22;

4. The departure of the weaving department manager from his job on August 1;

5. On August 12 and 13 the Lodz tramworkers' strike, which had some repercussions in the weaving department;

6. The loss by the section foreman of a sum of money and the organized collection in his behalf on August 15;

7. Notification of release of the section foreman on August 20;

8. Voluntary overtime work on Sunday, August 25;

9. The first meeting of the presidium of the workers' council to discuss the 13th wage on August 23;

10. The poster and the problem of workers' involvement on August 29;

11. The second meeting of the presidium to discuss the 13th wage on August 30;

12. The assembly meeting of the workers' council, also held on August 30, to consider the problem of the 13th wage;

13. Release from his job of a foreman who refused to sign an appeal asking tram employees to return to work, first reported on August 24.

1. The Weaving Department Production Conference:

Production conferences had existed before the October change, with attendance occasionally compulsory. It was reported that the section foreman toward the last had stood at the factory entrance, trying to induce his people to go to a conference instead of going home. I-10: "Before, they were pushing us to meetings."

Production meetings previously had been organized by the labor union, which helped to increase the overlapping of union and management in the minds of the people. But since October, 1956, they were held under the auspices of the workers' council.

The first production meeting we could attend was a meeting of the weaving department workers' council on July 17, starting at the time when the morning shift was being relieved by the afternoon shift, that is, around 2:00 P.M. Since this was only a departmental meeting, the small meeting room near the party secretary's office was used. The room could hold approximately a hundred persons. On the wall behind the chairman's table hung a great portrait of Gomułka.

Chairman of the meeting was the manager of the weaving department; he was not, however, a member of the departmental workers' council.[1] The members of the council sat around him behind the chairman's table, but during the whole meeting they did not take the floor once. With them sat the shift supervisor. Thus, the workers' council members were interspersed with management people.

The meeting was opened to all who were interested. Of our two work groups who were just beginning their afternoon shift, no one, with the exception of the section foreman and the shift supervisor, was attending. Of other persons already introduced, there were the supervisor of the bureau in his function of the chairman of the weaving department's workers' council, the departmental party secretary, and the labor union secretary. The plant director joined the meeting later, taking his seat at the side of the manager. Neither the party secretary nor the labor union secretary spoke during the meeting. Though formally organized by the workers' council, the meeting actually was run by the manager and the plant director. During the hour-and-a-half-long meeting, persons came and left.

The meeting was opened by the manager, who surveyed the production of the weaving department in terms of percentages of the quantitative plan. "As you know, we have met to survey what we have been doing during the last half year. We know that the weaving department has many problems. Now, the plan for the half year was 99 percent fulfilled. If we break it down, we find that shifts varied in their fulfillment of the plan. For example, the shift of shift supervisor X fulfilled the plan by 93.5 percent, while the shift of shift supervisor Y [this was our shift] met the plan by only 91.5 percent, and so forth."

The manager said that our section foreman had met the plan by only 84 percent. However, with allowance made for

[1] The manager was a member of the presidium of the workers' council. He was not supposed to act in that capacity on the departmental level, where he was primarily representing the management as a manager of the whole department.

interruptions due to reasons beyond his control, he had achieved 96 percent of his quota. If the interruptions were added to the total production of Work Group I, the group achieved 98 percent of the plan. On the whole, the manager said, technological factors of production were relatively under control, but what really mattered were psychological attitudes toward work. The latter were believed to be wrong.

It was also stated that 73 foremen had met the production plan, while 44 had not done so. With allowances for interruptions beyond their control, the number of foremen who failed to meet the plan was reduced to 30, about 25 percent of the total number. The qualitative plan, too, was not fulfilled 100 percent. There were too many wrong "draws," uneven weaving, and oil spots.

Having finished his survey, the manager asked the group to comment. Several persons whom we did not know rose to explain the failures. Some of the comments were offered in self-defense. There was a tendency to shift the responsibility to other persons or to factors which were beyond the control of the speaker. For example, the first complaint referred to the roofs. Just that day, a heavy rain was falling, and leaking water was dropping down on the looms. In several parts of the shop, workers protected their looms by stretching canvases over them.

The manager said that to fix the roof required a major repair that might amount up to fifty million zlotys. Such a sum could not be appropriated by the plant without the consent of the ministry in Warsaw. Obviously, such consent would not be easy to obtain. In other words, it was felt that this issue was beyond the control of the management, at least at that time.

Another complaint was that the morning shift sometimes was lacking a sufficient amount of filling. The reason was quite simple: The filling winding department worked only during the day shift and sometimes did not prepare a sufficient supply for those who started at 6:00 A.M.

Other complaints: The temperature was very high in July; people could not work as well as usual. Some persons are not able to handle six looms at once; such persons should attend no more than two looms. Poor cooperation between foremen from different shifts. This last point was important especially for Work Group I. It was said that the foreman who alternated with our Foreman I had a habit of leaving the looms in a poor state. Thus, Foreman I was always very busy at the beginning of his shift, repairing looms left in bad shape by the shift that had just left.

While particular complaints were pouring in, the plant director walked in and immediately took part in the discussion. At that moment a woman complained that a frequent change of the kind of material to be produced sometimes required a stoppage of production for the whole morning, causing a delay in adjustment and resetting of harnesses, and so forth. The plant director replied that such a change should take no more than an hour. As he said this, workers who were sitting around us ridiculed his statements in subdued comments.

Our shift supervisor then lectured on discipline. Next, our section foreman said: "What do we hear here? Foremen maintain that weavers are guilty. Weavers maintain that foremen are guilty. But real guilt can be found in the material, this cotton we have to use." And he held up a piece of cotton.

This comment did not please the plant director. He maintained that the weaving department was sufficiently supplied with all technological items. And he also promised that if there had been some deficiencies in the past, there would be none in the future. The plant director maintained that weavers should report foremen who do not do their jobs properly. Presumably, the plant director had received several complaints from workers that the foremen (since their new pay regulations) did not care as much about work as they did before.

"Complaints are," continued the plant director, "that sometimes it was necessary to look for the foremen even in the foremen's dressing room where they are resting. Thus, weavers ought to tell which foremen fail to do their duty."

The plant director's direct appeal produced a response from one woman, who said that her foreman ought to repair looms faster. However, two other women from the same work group defended him.

At the conclusion of this meeting another series of complaints was presented. Our section foreman listed three factors: low wages—some women were making only between 800 to 900 zlotys; dirt on and around the looms; lack of reserve foremen and helpers.

At the very end, there was a direct attack on the manager by the plant director. The plant director said that since all technological factors were satisfactory, the responsibility for the failure was to be fixed upon the management that did not succeed in properly organizing work.

In response, the manager said: "We can observe a clear tendency to shift and project the responsibility from oneself onto other persons. However, everybody is responsible for his part played in the production process. Therefore, a worker in the management position is responsible for major and essential issues and not for minor failures."

A review of the production meeting discloses that complaints were frequently expressed, but no suggestion for their removal was offered by the management. The plant director denied the existence of technical problems, yet we learned later that technological features have been improved.

2.   The Foremen's Production Conference:

The second production meeting, held in the afternoon of July 27 in the great assembly hall, was open to the whole plant, but was a meeting for foremen only. Since our groups were not working the afternoon shift that week, both foremen could have been present; however, only Foreman II came. Of course, section foremen, shift supervisors, and all other management persons were present.

Since this large room seated about five hundred persons, the long speaker's table was elevated. Behind it, in addition to the chairman of the presidium of the workers' council, were sitting the plant director, the plant party secretary, and the plant labor union secretary. Thus, representatives of all four organizations faced some one hundred persons. Behind the chairman's desk was a great weaving trade emblem and a poster showing an outline of a man walking ahead of a mass of people; the caption in great red letters read, "The Party."

Bottled soda water was offered to all. Everyone at the chairman's desk had an opened bottle before him.

The meeting was opened by the chairman of the workers' council presidium with a roll call. Since a number of foremen were absent, he stopped reading, said a few preliminary words, and then introduced the director. The director walked to the microphone and read his long report about the production situation in the plant. Several of his listeners appeared quite bored.

The plant director started with a general survey of the economic situation in the country during the first half of 1957. He said that production had increased by 9.4 percent, but this was mostly in consumption goods. It was planned that incomes should be increased in the third quarter of the year by 6 percent. Therefore, it was necessary to supply more goods to the market. Since there was enough raw material available, there was no reason why the plan could not be fulfilled. And yet, twenty weaving shops in the northern district of the cotton industry had failed to meet quotas. The whole cotton industry during the first half year was lagging with a deficit of two million meters of cloth.

The director said that the plant produced goods valued at 560,336,900 zlotys, 103.2 percent of the quota of 542,895,500 zlotys. However, he said there were unnecessary losses in several departments. One of the major problems was absenteeism. During the first half of 1956 the absence rate was 4.1 percent, and only 0.4 percent was not excused. For the same period in 1957, there was 5.47 percent absenteeism, 1.73

percent of which was not excused. A tightening of discipline
was urgently required.

After reviewing and criticizing production in several other
departments, the plant director turned to the weaving department, which he said had failed to meet the plan by 0.2 percent.
He blamed the failure on lack of efficiency in organizing
overtime Sunday work and on the 4.32 percent absenteeism,
which caused 433,897 meters of material to be lost. Who was
responsible?

"Responsibility rests on the manager, the section foremen,
and foremen. For example, the section foreman [the plant
director named our section foreman] does not control the
situation organizationally. The greatest number of trainees
have been leaving this section foreman because he was not
able to provide for them the necessary and satisfactory conditions for work."

The plant director criticized section foremen for their concentration on checking absences and not caring about technological processes of production. He also accused them of
carelessness about yarn and other supplies. While mentioning
two foremen who went over their quotas, the plant director
also named two foremen who were producing under 90 percent.
The plant director again criticized the manager: "It is re-
grettable that the management of the weaving shop does not
organize production conferences for foremen only. In such
conferences, foremen could share their experiences, and this
would contribute to the development of collective spirit, and
through that, it would increase production."

The director pointed out that the qualitative plan also had
not been fulfilled. He said, for example, that Loom Number
1253 while producing material identified as "BP-70" had had
five successive mistakes. "And what would be the answer of
the section foreman?" asked the director. Since the question
was directed to the young section foreman, we loked around
and saw on his face a faint, embarrassed smile.

After specifying mistakes by particular persons, the director
listed the losses due to rejects, which had amounted to

18,914 kilograms of yarn during the first half year. He also said there was too much dirt around the looms.

In conclusion, he appealed to foremen: "Comrades, it is necessary to increase our attention to work. If all of us—weavers, foremen, section foremen, and management—unite in a collective effort, the plan can be reached."

When the director finished, there was silence. Finally, the chairman of the workers' council said that there would be a discussion. "Who would like to say something?" No one raised his hand. The foremen sat silently, watching the chairman. Suddenly, one rose, but he just walked over to the chairman's desk to pick up a bottle of soda water. The plant director gestured as if to say: "That's for you. All of you take a bottle."

Once more the chairman invited comment. Then the manager of the spinning department walked to the microphone. He spoke in defense of the foremen. He said that the trouble was that there were no rules of work discipline. The foremen were not responsible for slackening of morale. When he had finished, the chairman asked the foremen who were members of the workers' council to speak out. However, some foremen walked out of the assembly hall.

The next speaker, the manager of the printing department, said that the majority were good workers, but a few sometimes exerted a bad influence upon the others. He said that more apprentices were needed.

The secretary of the plant labor union then mentioned that not even one department had showed an improvement. And yet, he said, the 13th wage depended upon the improvement. He stressed the need for changing attitudes toward work, toward the factory, and toward Poland.

The plant party organization secretary, who spoke next, said that the duty of party members was to be everywhere first, to accept greater responsibilities. He asked that the party members seek as much contact with the rank and file as possible. He appealed to the foremen's honor and patriotic feelings in combating absenteeism.

After these, several managers and shift supervisors gave

short speeches, mostly reading from prepared notes. Most of these statements were exhortative. One foreman, presumably a party member, elaborated on the purpose of production conferences, to exchange experiences. "How can we know what is wrong if we do not talk to people?"

While these short speeches were following each other in quick succession without really starting a discussion, the plant director rested behind the desk, munching a sandwich. When the problem of the purpose of the meeting was brought up, he again took the floor. This time he spoke without notes. He was a good speaker. He criticized the foremen because only one foreman took the floor. He would prefer a foreman who did not fulfill the plan and yet came to the meetings and spoke out, to a foreman who fulfilled the plan and did not come to the meetings. In full swing, he again criticized the manager of the weaving department, maintaining that all conditions were favorable and yet the plan was not being met. "I am asking why!" shouted the director, waving his hand. If the foremen lack discipline, the weavers will, also, he said. And whose responsibility is it? The manager is responsible because he does not use strong measures against foremen who fail to meet the requirements. "However, if we start working together," said the director in conclusion, "we shall overcome the difficulties. I appeal to your professional honor, to your honor as citizens and Polish patriots. Let us get up together and start again with full power. I promise you on the part of the management that we shall do everything that is necessary. Let us again become the first textile factory in the country." This dramatic appeal drew some applause from the audience.

The meeting was almost over when someone from the audience raised the question of the two foreman delegates to the all-textile industry labor union meeting, in which there was to be a special foreman section. Our shift supervisor suggested two names. The weaving department party secretary opposed, stating that the candidates ought to be discussed. But there was no reaction on the part of the foremen. The nomination of the two foremen was accepted, and the chairman of the workers' council closed the meeting.

In the following days we found out that the speech of the director had not been received favorably.

The shift supervisor said, "Several foremen left. They do not like it. A conference ought to be a discussion."

Foreman I: "I was not there. I heard that he again shouted."

Foreman II: "I left after a while."

The young section foreman: "You see how he shouted. He starts to do it again the same as he did before October."

Interestingly, the weaving department party secretary also freely expressed criticism: "I believe that the plant director was wrong in stating that he liked a foreman better who does not fullfill the plan and yet comes to the meeting, than a foreman who fulfills the plan but does not come to the meeting." And concerning the election of the delegates, the secretary maintained that this was not a proper way to elect a representative. He said that the foremen had been overrun, that they ought to have had a little more time to think it over. The departmental party secretary used in this connection the term "undemocratic." Nevertheless, both Foremen I and II felt that the delegate, whom they knew personally, was a good man. However, Foreman I said: "You saw the election of our representative. Somebody proposes a person. One does not have time to talk it over. They always organize in such a way. . . . Why the people did not talk? People are afraid. If you talk too much, you might have troubles."

3.   The National Holiday of July 22:

On the eve of July 22, the Independence Day of the Polish People's Republic was observed by the plant. The meeting, at 5 P.M. in the "House of Culture," a red-brick building opposite the plant, was opened by a speech of the plant party secretary on the thirteenth anniversary of the People's Republic. Behind a table on the podium sat, in addition to the plant director and other plant officials, an officer of the Polish army. Most noticeable was the absence of all formalities. The secretary wore a tie, but the plant director did not, although his shirt was buttoned; the shirt collar of a young man sitting behind the chairman's desk was open. While the ceremony was going on, people kept entering and sitting on creaking chairs. One man, bringing with him two boys, came in a T shirt. When my coworker asked him to put on his jacket, he answered, "Does it matter?" and continued sitting without it.

While the party secretary was giving the major address, the little table lamp started to smoke. Nevertheless, he unflinchingly kept on. He stressed achievements of the Polish People's Republic and the great sacrifices the Polish nation had contributed to the Allied victory in World War II. He listed not only eastern battlefields, but also Tobruk in Africa and Monte Cassino in Italy, where the Polish Western Army gave a good account of itself by suffering bloody losses. Finally, he stressed the friendship between Poland and the Soviet Union.

After the speech, prizes and honorary crosses were given to persons who had excelled at their work. A supervisor was awarded a silver cross, and the citation signed by Premier Cyrankiewicz was read. Several bronze crosses were awarded to other persons. Finally, all those who had completed twenty-five years, thirty-five years, and forty-five years of work in the plant were given monetary rewards. Several of these older persons were absent and others did not hear their names, causing a mild confusion. The recipients of the awards were congratulated by the plant director and the chairman of the workers' council presidium.

Entertainment followed. Most of the numbers were provided by professional artists from Lodz theaters or broadcasting stations; only one skit was danced by a group composed of workers themselves.[2] The audience reacted feebly to the per-

---

[2] Some years ago, there had been an amateur theater and several different artistic groups, led by a professional actor paid by the factory. However, gradually these activities died out. In our studied group, to a question asking whether the respondents once in a while went to the House of Culture, three persons answered "yes," twenty-one persons "no."

formances. Several jokes, criticizing failures of contemporary social and political life in Poland, were probably not understood by the audience.

Reflecting on the ceremony, I would say that there was a lack of enthusiasm and "sacredness." However, the attendance was free and the hall was filled to two-thirds of its capacity. But, it was again the management, the party, and the union that were organizing the whole show, and the mass of the rank and file were passively attending. From our group nobody, with the exception of the foremen, attended.

4.   Departure of the Manager of the Weaving Department:

On August 1, during the afternoon shift, there was an informal meeting in the dressing room of the foremen, attended by the plant director, the manager of the weaving department, the party secretary, and the labor union secretary.

While some men were changing their clothes, the manager opened the meeting: "As you know, I have been transferred. I would like to thank you very much for your cooperation. I am unhappy that I have to leave now, when the production of the weaving department is a little low. But I hope that you will keep in mind your professional honor. I trust that you will again raise production. My position will be taken over by a new incoming manager. Would you please give him your loyalty and support as you gave it to me? And, once more, thanks to all of you."

The plant director replied: "Comrade manager should be congratulated because he has been advanced. He has become plant director in another factory. We should wish him the best. In that plant the production situation is even worse than in our factory."

Then the plant director shook hands with the manager and the manager started shaking hands with all present. The shift supervisor said: "In the name of the whole weaving department, I would like to thank the manager for everything." And other foremen came and thanked him.

Later, in the office of the manager, there was another meeting, attended by the section foremen and shift supervisors, and the same management people as before. This time the manager spoke mostly. In his calm voice he briefly reviewed the troubles the weaving department was experiencing. He again asked those present to give all possible support to the new manager, and then went around the room and shook hands with all present, embracing the labor union secretary. The lady wept. The plant director said: "We give you our best wishes in your new position. As a matter of fact, we shall be competing." The director giggled and shook his finger at the manager.

Later, when talking to people, we discovered that they felt that the plant director had won his struggle against the manager. I-5 said: "I would like to find out why he was moved to the other factory. You know, it looks like a promotion. But, sometimes, when somebody wants to get rid of somebody, one 'promotes' him."

On all these public occasions, most people did not address each other as "comrade" but by the professional status label, attaching to it *pan*: thus, "Mister Supervisor," "Mister Manager." The plant director and the party secretary were the only men whom we could hear occasionally using the term "comrade."

5. The Strike of Tramworkers:

On August 12 and 13 the Lodz tramworkers went on strike because their demand for a wage raise was refused. While Lodz was paralyzed, the army supplied trucks that transported people to work. To break the strike, the police allegedly used clubs and tear gas. All three Lodz dailies attacked the strike in front-page editorials and urged the employees to resume work. On Wednesday, August 14, the trams were running again.

What was the reaction of our workers to the tram strike? Ten persons volunteered their opinions:

Foreman I: "This was a mistake because the strike undermines Gomułka. It plays into the hands of the Natolin.[3] However, the police ought not to have beaten them. . . . Anyhow, they were not united. One-half wanted to continue work; one-half did not."

II-32: "The police defeated workers. But the workers did not really lose. . . . I took part in strikes in 1933 and 1937. I was jailed for five weeks because of that. But, at that time, all factories stopped, streetcars were turned over. Several labor unions united to carry out the strike."

II-15: "Workers want more bread, but they will not get anything more. All tramworkers in all Poland can get only fifty million zlotys."

II-17: "The strike is illegal. They say that the strike is illegal. Do you see?"

II-13: "Before the war, tram and railway workers had their own insurance society, they were relatively well off. Now, everything has been taken away from them. And this creates great dissatisfaction."

I-7: "I don't think they were clever in organizing the strike. What can they achieve if they go alone, just the tram employees? At least the whole of Lodz should have gone on strike. Furthermore, they only get between eight hundred and fourteen hundred zlotys as we do. If they would get a raise as a result of the strike, all other factories would have to do the same."

II-24: "Oh, I believe the tramworkers ought to get higher pay. However, compared to us, they have it even better. Look, they get free uniforms, free shoes. It is true that in the morning and in the evening there is a crowd on trams. But otherwise during the day, they just keep rolling around while we have to stay on our feet all day long."

I-4: "Oh, they were beaten. I know from some doctors in the hospital that even women had to be treated for injuries."

The reactions of our workers were not uniform. Although

[3] The Stalinist group within the Polish Communist Party.

all felt that physical force should not be used against workers, when they compared the tramworkers' demands with their own situation, their solidarity was dispersed. Some respondents realized that the odds against the tramworkers were too geat to give them a chance of winning the strike; from this realization resulted a sort of collective frustration.

The echo of the strike in the plant was, however, not yet finished.

6.   The Section Foreman Loses His Money:

An interesting example of spontaneous collective action was demonstrated in connection with the section foreman's loss of his midmonthly pay advance on August 15. According to I-5, the collection developed as follows: "I-14 came to see me and suggested that we should collect for the section foreman. You know, she is completely without any political understanding, and yet she proposed it. She said that each of us should give 10 zlotys. I told her to wait a little, that I would find out."

The section foreman himself reported: "I got my 1,000 zlotys and lost it. I know that I-14 suggested the collection to I-5. All eight work groups participated, everybody giving 10 zlotys. They collected 970 zlotys. Later, I found the money. Thus, I asked the woman delegate and II-34 to return everybody's money with my thanks."

Foreman I: "You see, before October they would not give anything to help him. Now the distance between party members and other folks is not as great as it was before. The party members do not keep their heads high any more."

I-5 said: "There is solidarity in misfortune. If somebody dies, we collect for a wreath. But there is no solidarity in production. We were not interested in raising it. Before, we came to work overtime only once. And this was because he personally appealed to people to do so; otherwise, they would not show up."

Since the problem of motivation for the collection was of

interest, we asked all persons the question, "Why did you contribute to the collection for the section foreman?" The explanation was rather uniform. Only a few replies are quoted here.

I-12: "He is not so bad. It was necessary to help him. He has a wife and children."

I-6: "He is a good fellow. He completely lost his head. He has a wife and children and no property. What would he eat? For us to give 10 zlotys each is nothing. Foreman I came to see me and said that the section foreman had tears in his eyes. He said that the foremen would each contribute 50 zlotys and the weavers each as much as he could. But some figured out that if each gave 10 zlotys, this would be enough. In our group, the helper collected the contributions. Everybody gave their 10 zlotys eagerly."

I-10: "It was impossible to let the children go hungry. Everybody contributed because it was not that much."

II-20: "He lost his money. How could he go home without money? What would his wife say? Everybody contributed a little."

II-22: "We felt sorry for him. What would he live on? He had a wife and three children to support. Ten zlotys for us was nothing, and for him altogether, the amount meant something."

I-8: "He suffered a loss. It was necessary to help him. He has a wife and three children."

Now, these were the responses of women. The major motivation was to help his wife and children. Obviously, it was very easy for the women to identify themselves with the section foreman's wife. Now, turning to the men, we get the following answers:

II-17: "People could not see an injustice. He has a wife and children. We always are one, helping out each other."

II-3: "The foreman said that the section foreman had lost his money. Everybody contributed 10 zlotys."

Foreman I: "He looked so broken, we felt pity for him. We

collected the money and gave it to him. He has a wife and children to support."

I-7: "I did not know that there was a collection. I did not give anything." However, according to I-5, I-7 knew about it and presumably had said: "He is better off than I am anyhow. I shall not give him anything."

II-15: "Well, the people were led by decency. Anybody would feel hurt, were he to lose his money."

The motivation listed by the men was similar to that of the women, but not as uniform. Checking with the young woman from the parallel work section, we got the same responses in terms of needs of wife and children.

The collection money case demonstrated that the workers could act on their own if the need arose. It should be noted, however, that they did not use a new organizational structure and that the action was simple. The news was disseminated by helpers and foremen. Helpers also collected the money. In addition, the immediate authority of the foreman in supporting the action was helpful. Thus, the "management-production" structure once again was used to bring about the collection of money. While the management often appropriated to its purpose other plant organizations, here and in other instances the workers and these organizations used some parts of the management's structure to their ends.

The collection of money also showed that there was not a great social distance between the workers and the section foreman, who was considered by some as belonging to "them," the management. We shall review the problem of the relationship between the workers and their immediate supervisors below.

7. The Section Foreman Is Fired:

As soon as the problem of the lost money was settled, another issue appeared. The section foreman was told that he would be transferred to another factory.

In this connection it is necessary to record that at that time a new manager of the weaving shop was already present. We saw him walking with the plant director, who was showing him around. Since he seemed to be busy and naturally not familiar with the new job and its problems, we did not contact him until on August 21, asking him about the reported dismissal of the section foreman.

The new manager: "Yesterday there was a meeting organized by the plant director and attended by the departmental party secretary, the labor union secretary, the shift supervisor, and myself. The section foreman was relieved of his job because the plant director was not satisfied with his achievement. Presumably, the section foreman himself had already asked for the release. He does not seem to care too much about his job. Anyhow, I am completely new on the job; I really don't know the background of the whole story."

The section foreman was upset. When I met him, he refused to talk to me, stating that he would never trust anybody. Foreman I told me later that he was suspicious that my colleague or I had reported him to the plant director. However, the next day, the section foreman smiled when we showed up and talked to us again in a friendly way. He said he was sorry for what had happened the day before. However, he knew that we had nothing to do with his dismissal. Later, Foreman I told me that he had assured the section foreman that it was not our fault.

The section foreman maintained that the plant director had a grudge against him: "He knows that I have more education, and he is after me. When the manager was fired, it was then my turn." He said that the manager had told him privately that the plant director had maneuvered the manager out of his job; naturally, the people could not be told about it because they would dislike it and eventually make trouble. The section foreman felt that the plant director might think it over before firing him if he knew that people would start trouble.

Later, the section foreman reported that somebody had presumably denounced him. The same was confirmed by the plant director, whom we asked about the case of the section foreman. The plant director maintained that it was reported that the section foreman had approved of the tramworkers' strike. Obviously, this was something one would not expect from a person in such a responsible position!

An older section foreman confirmed the above statement: "He said something approving the strike. However, he is a party member and he can defend himself. I am not a party member. This is a purely dictatorial decision of the plant director. He tries to remove all persons who have good relations with people. The plant director is afraid that such a section foreman could have good relations with foremen and weavers, and this they do not want."

According to the section foreman, there was not any chance that the labor union could or would help him. He did not believe that the party would step in on his behalf either. My colleague told me that since the section foreman was a member of the party, he was supposed to oppose the strike. There was no chance that the party would back him; just the opposite could be expected.

While discussing the case, it was brought out that there was one more body to which the section foreman could appeal. There was a committee of revisions, to which one could appeal if the party, the management, or the labor union were unable to handle a case satisfactorily. However, the section foreman rejected all the suggestions, maintaining that they were of no avail.

After the weaving department party secretary returned from his vacation on August 31, there was a meeting of the party departmental organization. The section foreman knew of it, and I-5 and Foreman I also knew about it. Encouraged by Foreman I, I-5 decided to go out to help the section foreman. He assembled a group of workers. With the woman delegate, II-34, and the young man from Work Group III, he went to the party meeting. The section foreman faintly opposed it, but I-5 went ahead with his idea.

I-5: "I spoke, asking them whether they would admit us. I asked them kindly not to remove the section foreman. I said that he had our full confidence, and so forth. The plant director answered that he was pleased to hear it. He also said that he was favorably impressed by the more than 100 percent participation in Sunday overtime work under the section foreman's supervision. He also said that the matter would be positively settled."

Thus, the section foreman was saved. Later, on the night of my leaving, he reported to me that he had been promoted to shift supervisor.

The reader ought to note that the initiatory role of I-5 was important. Within the factory setup, the workers were supposed to send delegations to higher factory authorities. However, since the delegation did not go through the managerial hierarchy, but directly to the party, this contributed to a certain undermining of managerial prestige, at least in the weaving department.

There was one more interesting point. When I-5 was discussing his intention to intervene in behalf of the section foreman, he said that the management could not disregard feelings of workers. I-5 said: "They say that the workers rule in the factory. All right, let us see." Thus, a frequently ridiculed statement in this situation was used to enhance the position of the workers.

8. Sunday Overtime:

A few days before, while we were sitting in the smoking room, the section foreman came in and spoke to the men: "We shall have overtime work on Sunday. I personally beg you to come. I am this time personally very much interested that you should come. I also asked the foremen and other persons to do it as a personal favor to me."

When he walked out, I-5 and the young man from Work Group III said that they would come. They felt that when they were asked in such a way, it was obviously proper to come. The women felt the same way.

Commenting on overtime work in general, Foreman I said: "Before October, when the plant management wanted the people to come for overtime work, they would not come. It was necessary to ask them personally. They came because of me. I remember that once the management wanted to set a good example and they themselves worked on the Sunday brigade. We were pleased to see that the plant director was poor in his work. Of course, immediately after October, the situation changed; people were coming by themselves. But today the people are again less willing to work overtime. They say that it is beginning in the same way as before. Well, the management has decided and we have to come. We shall work from midnight to six o'clock Sunday morning. You see, the workers' council did not have much to say about it."

Foreman II was on vacation, being replaced by another foreman, a bulky man with a bald head and a habit of talking fast. The replacement foreman ran around talking to the people in a somewhat direct approach: "I saw every weaver and told him: 'You will earn more; you will get more bread, new shoes, and so forth. Why not be clever and come?' "

The overtime was a success. The section foreman was beaming. Just a few persons did not show up. Within Work Group I, I-7 did not come because of his sick wife. II-17 had said that he might not come, and he did not. II-20 also had said that he would not come, presumably because of a bad headache. The section foreman told me gleefully that all persons who promised to come did show up. He also reported to me that, for example, II-9 earned that night 74 zlotys after all deductions had been taken from his paycheck. A highly distrustful attitude was shown by I-7, who did not come. He said: "My monthly plan has already been fulfilled. I would earn some thirty to forty zlotys on that night; they would not

give me more. They would figure it out somethow not to give
me more. So, it is better to stay at home."

The suspicion of I-7 was refuted by I-5 and Foreman I.
According to them, the weaver was paid his normal piece rate
plus 4.75 zlotys for every overtime hour. And this was practically a 100-percent increase over the normal pay.

In connection with the overtime work, the young man from
Work Group III expressed a complaint: "There is a poster in
the shop, signed by the Minister of Light Industry. There it
is stated that no overtime work can be required of workers.
And yet, they ask us to work. And it is like that all the time;
one says something and one does something else."

The ministerial statement Number 324, issued in November
1955, was in fact posted at the side of the main entrance.
When I asked the section foreman about it, he replied that no
worker was forced to come. This time the Sunday work was
really voluntary.

As Foreman I reported, the immediate supervisors of the
weavers had developed their own way of getting some unpopular demands across to workers. They placed it on a personal
basis. There was no element of collective or national goal.
While the plant director in his speeches manipulated the
national honor in order to increase readiness to work, foremen
and section foremen presented it in terms of personal appeal.
This made it possible for workers to identify themselves with
the foreman or section foreman who had the bad luck of having
such a position that they were responsible for the execution
of unpopular tasks. One could understand it and, therefore,
one could help such a fellow if he was "a good chap."[4]

The second point observed in this connection was the dislike

---

[4] I-6: "You know, my dear sir, before, they were pushing us. Those who
worked above the norm got a premium. They were photographed and their
pictures were posted on the billboard. We did not like it. There was a woman
who walked around and kept repeating: 'Work more, work more.' Foreman I
was also pushing us. They frequently called him to the office and asked him
about the percentages. Oh, he had quite a lot of troubles. He came and
asked us: 'Jesus-Marie, I beg you to help me. Believe me, I am not a member
of the party. Please, do it because of me.'"

of the management's reaching a decision without asking labor. When the ministerial statement was brought up in the smoking room, six men were sitting there. As the young man from Work Group III maintained that the management disregarded the decree of the minister, I asked whether it would not be worthwhile to write to the minister and ask him for the correct interpretation of the decree. This question evoked rather strong reactions on the part of our men.

Foreman I: "The best thing is to keep quiet and work."

I-7: "Yes, this is the right thing to do; otherwise they fire you. You might have difficulty getting another job. They might telephone each other about you, and you will not be accepted."

II-17: "I knew one fine fellow who was fired who could not find another job because no one wanted him. Thus, he had to come back and beg for reinstatement."

I-5 usually calmed his men down, trying to argue out some of the points. This time he, too, succumbed, and burst out: "They stick together; I can see it in the workers' council. They are always prepared; they know what they would propose. And that goes as quickly as that." After a short pause, I-5 added: "Before you can think it over, they have passed another decision. The trouble is that the foremen and all other management people are afraid to speak out. Only workers are not afraid. Only workers can do something. If they fire you, a worker is not afraid to pick up another job, but the management people are."

This was naturally a statement somewhat in contrast to I-5's usual optimism. Nevertheless, I-5 likewise felt bitter, as the other men did at that moment in the smoking room. They felt that they were being manipulated by some invisible power, by "them," who were "cheating the workers."

Where did the men and women draw the line dividing labor from management? The reader will remember that during the workers' council election several persons stated that they voted for I-5 instead of Foreman I because I-5 was "one of us." This would suggest that the foreman was already "management." But, on the basis of our direct observation, this seemed not to be the case. The foremen participated in all discussions workers had, worked physically as they did, were covered with sweat and dirt like anyone else. The line of division was drawn rather with the section foreman, who spent much of his time in the front office, worked manually only in rare cases, kept a tidy and clean appearance. According to Foreman I, there had been a change in the workers' relationship toward our section foreman: "Before [October], they would not help the section foreman if he asked them to work. The distance between him and the people was greater than it is today."

If the men accepted the section foreman as "management," I-16, the young married woman, disagreed: "The section foreman is not yet management. The management starts with the shift supervisor. The section foreman shouts sometimes. But I can argue with him if I want to go home earlier. You know, he has a timetable and knows when my train is actually leaving. But, nevertheless, sometimes he lets me go earlier."

Unfortunately, on this point we did not succeed in collecting systematic data. The section foreman was most likely a "marginal man" because he was meeting our men and women daily, to some degree participating in their troubles, while simultaneously spending some of his time in communication with the upper levels of management. Though the men considered him as management, the whole group was willing to act in his behalf, to identify themselves with him in his troubles. Therefore, the most acceptable explanation would be that the division line was not sharply drawn if frequent personal interaction was involved.

On the whole, management included persons from the upper echelon with whom our men and women had little to do. The more they were removed from the shop, the greater probably was the distrust on the part of workers.[5] And management meant especially those persons who faced our men and women at different meetings at the front of the room behind the desk. These were representatives not only of the plant managerial staff but also of the labor union, party, and increasingly more so, the workers' council leaders.

9. The 13th Wage Presidium Meeting:

During the month of August, the Lodz newspapers published several articles reporting the factories within the industrial district which had earned a 13th wage. This raised some hopes in the plant.

We were informed by I-5 that there would be a meeting of the presidium of the workers' council. We were permitted to attend it while the 13th wage was being discussed.

The meeting was held in the office of the chairman of the workers' council (the chairman was on vacation). A young supervisor, the vice chairman of the workers' council, presided. The meeting was opened by a report of the director of business administration, who was worried because the plant fund from which one was supposed to pay out the 13th wage was very low—some 619,050 zlotys at that date. This would mean that each employee would not get more than 67 zlotys. Therefore, the director proposed not to pay anything, because it was so little that it would presumably create an unfavorable impression upon workers.

The vice chairman considered another solution: Since the finishing department had fulfilled the plan, it could be rewarded. Too, some individual weavers were successful, and

[5] Possibly a good index of who belongs to the management or not could be based on attendance at the "Club of Rationalizators," whose social rooms were located in a building beside the House of Culture. Originally, the club was supposed to be a meeting place of all who had some new ideas. In practice, this was a management club. Most of our men and women did not even know about it. The section foreman did not attend it.

they ought to be rewarded for their contribution to production. However, in light of the former dissatisfaction of workers with premiums, the vice chairman wound up by supporting the proposal of the director of business administration.

A member of the presidium opposed the proposal. He said that the people would feel that it was better to get even that small amount now. By the end of the year they would be afraid the money might be gone. Furthermore, everyone knows that the 13th wage has been paid out in several other plants.[6] So the people will ask, why not in our factory?

The director of business administration moved that a committee be formed to explain the situation to the workers—that lack of the 13th wage was due to the failure to meet quotas. Another member proposed that the workers should be told that if all departments had worked as the finishing department had, there would have been the 13th wage. It could possibly be figured out by how much the other departments failed to contribute to the 13th wage.

However, there was still opinion favoring the payment of the 13th wage immediately. I-5 made himself heard, maintaining that his people would surely favor the payment now.

The vice chairman became tense and stated that no member of the workers' council could go along with irresponsible demands of the people.

I-5 replied that he was expressing the voice of the people and that this was his duty and right.

The vice chairman pulled back a little and conceded this point. Nevertheless, he said, there is no solution left other than that proposed by the director of the business administration. He said that the departmental workers' councils ought to explain to the people and convince them; one ought to communicate it over the factory radio.

I-5 maintained that one ought to take a vote on that issue.

[6] As was reported, for example, in the Lodz *Express Illustrowany*, August 23, 1957.

The vice chairman proposed a meeting of the whole workers' council. He again stressed the duty of the workers' council members to convince the people about the soundness of the proposal, to explain to them where the 13th wage was taken from, and so forth.

The secretary of the workers' council proposed a date for the meeting of the assembly of the council, and this meeting was concluded.

The vice chairman had a slightly condescending and authoritative attitude toward other persons. He was an intellectual meeting the workers, and probably also used to handling certain issues somewhat high-handedly from pre-October days. But I-5 was not afraid and replied.

Referring to the presidium's meetings, I-5 complained that there were only two genuine workers there. All other persons were at least foremen. I-5 felt isolated. "There are too many white-collar people, and few real workers. There ought to be at least one-half workers."[7]

When I-5 returned to the weaving shop, he met his men in the smoking room. As one could expect, all of them were in favor of distributing the money immediately. II-17 then told both young women, II-34 and II-32, about the issue. Both of them wanted to get the money immediately.

During the meeting of the presidium, one interesting conflict in roles appeared. When I-5 spoke, he spoke as a delegate of the people who elected him. When the vice chairman spoke, he defined the role of a member of the workers' council in terms of a transmission belt, of a helper of the management who accepts the management's point of view and puts it across to the electorate. The practice so far had been that all other channels had been assimilated by the management to its own purpose. The labor union and the party had been communicating management's demands down to workers.

[7] According to the workers' council law, two-thirds of the workers' council members were supposed to be "workers." However, in the bylaws of the workers' council, no such stipulation was stated. See the Appendix.

Now the workers' council was a new body and, therefore, there was an uncertainty where it would stand and what it really would be like.

10.   The Poster Incident:

A dramatic event occurred on August 29. The plant director posted a statement: "600,000 zlotys are sought for the 13th wage." And an arrow pointed to an open large box full of spoiled and rejected yarn. The poster and the box were placed close to the main entry.

While the morning shift was leaving, a crowd of workers gathered around. The crowd got angry, and the plant director came to calm them down. We observed that he was surrounded by people, mostly women, while other upper echelon people held themselves apart from the crowd. The new manager and the weaving department party secretary later told me that this was the plant director's idea.

Voices shouted: "Before, there was always some profit; why don't we have anything now? How is it that in other factories they get something and we do not get anything?" While the people talked to the plant director, they addressed him as "Mister Director." But a few steps further away, they used less polite words.

According to an eyewitness who attended the demonstration from the very beginning, this is what happened: "While I was coming to work, around 12:40 P.M., they were just posting the placard. The plant director stood aside. People stopped and started to ask: 'How come? No 13th wage? "They" stole and wasted the money, and now they want to find a scapegoat.' The new manager approached the plant director and asked: 'What are you doing? This is nonsense. You will stir up all the people.' And the people got very angry with the plant director and scowled at him. The plant director left for the factory radio room, but later he returned and admitted that the yarn was from the automatic weaving shop and not from our shop."[8]

By five o'clock that evening the box with the spoiled yarn had been removed. It was offensive to workers because it inferred that they were responsible for the fact that there would not be any 13th wage. Probably, if the management had said that there was only a small 13th wage and that all, the management included, were responsible for the loss of the money, the people would not have felt so bad about it.

Speaking later to the shift supervisor and Foreman I, we found that the shift supervisor was critical of the plant director's experiment. He said: "This was silly. The plant director himself formulated the idea. No other member of the workers' council knew about it. Somehow, the plant director rules along the old Stalin line. We cannot do anything. But this will be changed; the old Stalinist line will be removed."

I-5: "Yes, but why don't you say it directly, face-to-face, to the plant director? Why are the management people afraid?"

The small shift supervisor grinned self-consciously and said: "Well, you know, they have families, children . . ."

Later, I-5 commented: "You see, management people are afraid. I am not. Only the workers can do something." Though this was an overstatement, it was true that the management personnel were more cautious than the workers.

Another immediate result of the plant director's visual experiment, it was alleged, was a drop in production. The people also refused to work overtime. Said the shift supervisor: "This is bad. We have been catching up, and now the people say that they will not come for the overtime shift."

While such talk was going on, the group of our men met in the smoking room. Feeling was quite high. Almost all of our

8 There were two weaving shops. The automatic shop was located on the third and fourth floors in the adjoining building. There a person was in charge of eight or twelve looms, a recent Polish home production, which caused great dissatisfaction because they had serious deficiencies. The factory was seeking partially to rebuild the automatic looms. Our group worked in the mechanical weaving shop.

men were present. For the first time they seemed to be genuinely interested in the problem of management and financing. Despite the fact that the plant director's poster backfired, it aroused the minds of our men. The men felt that the poster's accusation was false because the waste of money was due basically to the mismanagement of larger sums.

The young man from Work Group III: "Why was so much money spent on the repair of roofs? And look, it is no good. It still leaks. How much was spent on the repair of the other building?"

Men expressed the desire to see the balance sheet. I-5, after a certain hesitation, promised that he would ask for the balance sheet at the next meeting of the workers' council presidium. Later, he changed his mind and said that that would be impossible because of the fact that the issue of the 13th wage would be discussed. Some men were, however, distrustful of I-5. Thus, I-5 said: "All of you can come and listen to what I shall say." I-5 felt quite bitter when telling me, "Only if Foreman I comes and helps me will I be able to fight against the one-sided opinions of our men. They look upon the problem only from their own limited viewpoint; the scope of the national economy is irrelevant to them."

Later, when the emotions receded and I talked to the young man from Work Group III at his looms, he self-critically admitted: "Well, the men are really not interested in the management issues. They get involved only if their pockets are concerned; otherwise, they do not."

Later, the supervisor of the business office stated in reference to the balance sheet: "We ourselves do not fully understand it. It is too complex."

This was a tone of resignation. The reader will remember that the supervisor of the business office also was the chairman of the weaving department workers' council. Speaking to another member of the departmental workers' council, I was told: "This is true. Our workers' council has not done anything. It would be good to go out and talk to the people. But this would require a certain dedication for which we do not have the time."

Analyzing the above statements, we can conclude that they point out two basic problems of the workers' council, namely, the problem of how to get the people more involved, and the problem of explaining concepts that seem complex to simple, poorly educated workers performing their duty at the rattling looms.

11.   The Second Meeting of the Workers' Council Presidium:

On the morning of August 30 there was a meeting of the presidium. The meeting was called at the last moment, obviously an emergency meeting to discuss something of sudden importance.

Invited also were the members of the weaving department workers' council, and our shift supervisor from the weaving department. There was also a person from the central headquarters of the textile industry.

The chief accountant opened the meeting by a report in which he ruefully admitted that the last figure concerning the plant fund from which the 13th wage ought to be paid was incorrect. A further computation had shown that there were only 250,000 zlotys available for the 13th wage, that is, that an individual employee would get no more than 26 zlotys. On this occasion, the chief accountant also elaborated on all three plans, that is, the plan of production, the plan of capital accumulation, and the plan of profits. He also mentioned that the weaving department had the poorest achievement in all respects. He surveyed other departments, pointing out that the finishing department had completed its plan by 104.5 percent.

The plant director later added that the weaving department had had 85 percent in excess of expected rejects.

I-5 defended the weaving department, maintaining that the large looms could not use the yarn entirely to its end quite well because of a technical problem. In addition, he said that 120 looms were left idle because there were not enough workers. He also called attention to the fact that the weaving department as a whole was burdened by the automatic shop, which had had trouble with the machines, meeting the plan only by 60 percent.

The plant director answered that there were only 60 idle looms at present and that the bad productivity was caused by bad yarn.

After some further exchange of opinions concerning the problem of production, the discussion turned to the 13th wage.

I-5: "Last time we spoke of 69 zlotys, but today about 25 to 20 zlotys as the amount of the individual 13th wage. How should I explain it to the people? How should I tell them that there will even be less than that small amount?"

The chief accountant: "Well, it is necessary to tell them that it is useless to divide such a small sum. If we improve our production during the second half-year, the individual share will amount to several hundred zlotys."

The plant director: "When discussing it, it is necessary to stress that the fund cannot be divided now, but that it will be at the end of the year."

The delegate from the finishing department: "People revolt. They say that the poster and the box of spoiled yarn shown to workers yesterday is an attempt to throw the responsibility upon workers. And yet, who is responsible for the plant? How about all the losses in investment, in repairing jobs?"

The vice president of the workers' council: "The trouble is that the workers' council does not function well in all departments. The members of the council fail to transmit the decision of the council to the people. In order to explain particular points, it is necessary to organize departmental meetings of the council. To oppose the wrong opinions of the workers, the representatives should be able to present effectively the ideas of the presidium."

I-5: "I hope I have the right to say here how the people feel."

The chairman of the weaving department workers' council: "I do not know anything about the issue of the 13th wage. And yet, people come and ask me. How can I explain it? Furthermore, the weaving department is working better and with a greater endeavor this year than in other years." He offered several figures to support his thesis. For example, the plan for productivity by one loom per one hour was 7.390 picks. The actual productivity was 7.474 picks, i.e., the plan of productivity was met by 101.1 percent.[9] Then he continued: "Our departmental council proposed to the plant director several ideas that were not accepted."

The vice chairman: "We have to raise the authority of the workers' council."

I-5: "We, the presidium, proposed a method to save the rest of the yarn. This was proposed by the former manager of the weaving department. Why did not the plant director do anything about it?"

The plant director: "This was within the jurisdiction of the weaving department manager himself."

The chairman of the weaving department council: "Well, so what, actually, are the rights of the departmental council?"

The vice chairman: "To propose motions to the presidium, but the presidium approves or rejects a motion."

The chairman of the weaving department council: "We sent in a long proposal and we have never received any answer."

The vice chairman, apparently unaware of this, looks toward the presidium secretary for an answer. She moves some papers. Obviously, the charge of the chairman of the weaving department workers' council is true. The vice chairman mumbles something.

[9] Later I obtained the following data about the productivity of the weaving department: When the weaving department manager took over the job in 1955, the plan was 7.196 picks per one loom in one hour. The realized production amounted to 7.621 picks, i.e., 100.9 percent. Though the production plan was increased meanwhile, it was realized this year by 101.1 percent. These figures cover merely the mechanical weaving shop. The plant director was held responsible for the purchase of the poorly performing looms in the new automatic weaving shop.

I-5: "How about our competence in the transfer of the former weaving department manager?"

The vice chairman reads a paragraph in the bylaws of the plant workers' council. It says that the plant director has a free hand in personnel matters. Then, the vice chairman surges to a counterattack: "It can be seen that some members of the presidium were not able to get across to their people the decision taken by the presidium here last time. This is very bad. And one must bear consequences."

I-5: "I repeat that I have the right to express the opinion of my department. This does not mean that this is necessarily my opinion. I ought not to be attacked because I present the opinions of the people. I present the views of the people. This is my right."

I-5 was very resolute and outspoken when making this statement. The vice chairman realized that he had gone too far. The other members of the presidium favored I-5. The vice chairman said something explanatory.

Another member of the presidium: "People ask about the roofs. How much money was wasted on it? And I do not know what to say."

The vice chairman: "This can be answered. Neither we nor the headquarters of the textile industry is responsible for it. There was a so-called bright man in the Ministry of Construction who forced us to get such roofs."

Since the hour had passed, the meeting had to be closed. It was pointed out that in the afternoon, at the assembly meeting of the whole workers' council, the issue of the unrealized 13th wage would be presented to the whole factory force.

12. The Workers' Council Assembly Meeting:

The afternoon meeting was again held in the great assembly hall. The vice chairman had at his side the second secretary of the plant party organization, a bald man in his early forties, and some other members of the presidium. The plant director and the plant party secretary sat in the first row of the audience this time.

The vice chairman opened the meeting. He read the report presented in the morning at the presidium meeting. The report gave the figures explaining why the plant fund was as small as it was.

While the vice chairman read, some persons came in, walking noisily. These were the workers who left their looms. We saw among them the woman delegate, II-34, II-32, II-17, and the young man from Work Group III. The section foreman came, too, but he sought to induce the people to leave the assembly hall. While this was going on, more people moved in.

The vice chairman raises his voice and says: "I beg all those who are supposed to work now to leave the assembly hall."

But a worker shouts back: "How come we cannot attend a meeting of the workers' council?"

The vice chairman replies, reading from the workers' council's bylaws: "According to paragraph nineteen, only those workers who have completed their duties can attend."

Angry voices answer the vice chairman's statement. The disorder becomes greater. Workers shout that they do not wish to be cheated any more, that they wish to attend the meeting. The vice chairman shouts: "Please, leave the assembly hall."

The second secretary of the party rises to his feet and shouts: "What does it mean? What does it mean? People, citizens, aren't you ashamed?"

No one pays attention. The turmoil continues. The vice chairman, the plant director, and other chief executives run to the exits, talking with individuals, persuading them to leave the hall. The shift supervisor pushes his people gently out the exit.

The invasion of the angry and dissatisfied workers was gone, and the meeting resumed. The review was quickly finished. The delegates were invited to join in the forthcoming debate.

There was silence. No one wished to speak. The vice chairman repeated the invitation. As there was still no answer from the audience, I-5 arose from his position behind the chairmanship table and walked over to the microphone.

I-5: "We have been criticizing the plant director quite a lot. But one cannot only criticize; one has also to propose some steps leading to improvement. Where are the critics? Why do they sit silent now when they ought to speak? I am asking you to speak out, to propose what should be done to increase the plant fund so that we can get a full 13th wage at the end of the year."

After a certain hesitation, the second delegate from the weaving department arose and delivered a speech full of questions: "Yesterday the weaving department was accused of wasting 600,000 zlotys. But who is responsible for the bad yarn?"

The second delegate showed several rolls of yarn in support of his statement that the responsibility lay with the spinning department, or even further, with the producer of the cotton. He continued: "They say that the weaving department is responsible for the lag in production. This is not true. The director is responsible. The party is responsible. The workers' council is responsible."

After another outburst, he shouts: "What is 28 zlotys? This is for a kid. And we have to pay for the blunders of other persons."

He gets a small applause and, somewhat exhausted after his oratorical *tour de force*, sinks back in his seat.

The vice chairman: "Well, we do not have the money. And if the money is not available, there is no use to talk about the past. It is necessary to consider the future, what course of action is to be taken now."

There is again silence. Thus I-5 again gets to the microphone and says: "Why do you not speak out? Well, it seems to me that there are several reasons. First, some people are afraid to speak. They say if they disagree they will be punished.

Secondly, there is a category of people who say 'to speak or not to speak, what is the difference?' But one should speak, and yet there are no such persons who would like to speak."

In the latter part of his speech, I-5 referred to the former weaving department manager's proposal that the unused tie yarn be deposited in boxes. For some reason this had not been done.

Another representative from the spinning department examined the tie yarn brought in by the weaving department people. He identified some as "ours" and "not ours." He explained that they had many young apprentices, and finally said: "We are very much concerned with it. The spinning department wants to help the weaving department. It is necessary to get mobilized so that we can do better in the second half of the year. It is necessary to have more communication among us, to have more consultations about how to improve production. If we work, if we have patience, we shall reach our goal."

Another member of the presidium added: "All of us are guilty. All of us are responsible for mistakes. Why do we not develop more sideline productions? The workers' council has collected many suggestions and practically nothing has been put into practice. Is this a good idea?"

The plant director: "I admit that it is possible that the workers' council has not been initiated into managerial activities. But if we start working together, we can succeed. Why did not the weaving department report the bad quality of the yarn? This is the responsibility of the weaving department. If the weaving department had notified the spinning department, quite a lot of troubles would have been avoided. The percent of seconds was 2.5 percent and we are making 3.5 percent now."

Having surveyed some other technical problems, the plant director continued: "It is necessary to approach the people in a human way, to talk to them, to find out about their problems. If all of us will pay sincere attention to problems, if all of us start first with ourselves, we shall get good results and earn more. Let us get on the job."

After the plant director's appeal, several persons from the management gave shorter statements calling for improvement and cooperation.

The second secretary of the party made the following statement shortly before adjournment of the meeting: "What kind of a conclusion can we draw from all these statements and speeches? We do not inform the people sufficiently. Is this true? This is true. It is necessary to present all our troubles at production meetings. If we cannot do it, the people will not talk. We evade certain issues. No one has attempted, so far, to explain certain issues to the people. People are naturally suspicious. They say: 'Instead of money, we get rejects.' And yet we cannot explain to them what the problems consist of. It is necessary to remove the differentiation between 'we' and 'they.' It is necessary to explain where the money comes from. If we are right, why be afraid to go and see the people and talk to them?"

The vice chairman of the workers' council closed the meeting by announcing that courses in business administration would be offered in the near future at the cost of 20 zlotys per person, and that production conferences would also be held.

An hour after the meeting, and when I-5 had returned to his job, we asked the people in our two work groups how much they knew about the meeting. With the exception of men who had met I-5 in the smoking room, no one knew about its results. Four persons, I-6, I-2, I-8, and I-4, knew that there had been a meeting, but others, mostly women, did not know even that.

Another strong, emotional attitude in connection with our inquiry concerning the degree of information about the meeting was expressed by the foreman from Work Group III: "No one in the morning shift knew that there would be a meeting.

You see, they do not want us to attend. They push us out of the assembly hall because they want to distribute the money just among themselves."

The afternoon meeting resulted, however, in one immediate improvement. I-5's initiative had brought an immediate installation of reject tie-yarn boxes. We observed that the workers immediately complied.

There was one more interesting feature on the part of workers. The men who knew about the negative decision on the 13th wage accepted it with a contemptuous gesture but without any further comments. Everyone understood that there was no chance to change it. On the basis of their previous experience, the workers believed that certain things simply had to be accepted—if bitterly, yet silently. Some persons had a ridiculing and sneering attitude toward the inefficient methods of the management. However, another reason for the compliance was also the relatively small amount of the possible 13th wage. The reduced 13th wage would amount to no more than an average luncheon for one person.

13.   The Case of the Fired Foreman:

Shortly before my departure, there was one more interesting event. A foreman in the weaving department was told he was fired.[10] Explaining why, he said: "They called me to the office. The new manager of the weaving department told me to sign a release statement. I asked why. He was evading the answer because, as he said, he was new here. Finally, the plant director, who was listening, dropped the word. He said that I did not behave properly during the tram strike. He said that I could have incited the people to go on strike too. I asked who had reported it. They said that the labor union secretary did, but she is on vacation now and I cannot talk to her."

The foreman was quite tense when talking about it. But

[10] From now on, this foreman will be identified as "the fired foreman."

he spoke coherently, articulating his ideas clearly. Later, I was informed by Foreman I that the fired foreman had belonged to a group of foremen who had sought to remove the plant director from his job in October, 1956.

The fired foreman continued: "Later, I signed the release. I shall be transferred to another factory. I am not afraid of work, but they are, the plant director is. I shall work in mines, if necessary; I shall clean trash, if necessary."

While the foreman talked in the foremen's dressing room, some other foremen gathered around and listened. There was a suggestion that the new manager of the weaving department ought to come and talk it over with all foremen present. Foreman I told us that the new manager refused to come.

The foreman quieted his colleagues: "I shall go even if they should reverse their decision. Please, colleagues, do not do anything, don't organize anything. I shall not get lost. My wife works at present."

One foreman said: "Next time another one of us can be hit." The hired foreman continued: "All right, I have to serve as a scapegoat. I shall take it, please."

The other foremen were not in a very belligerent mood. They advised our protagonist as to what kind of factory he should try to get into. Another foreman sneered: "They say there is a great need for foremen, and they fire him." Another voice added: "Because he knows too much, he knows too much about the plant director."

On September 3, when walking to the weaving shop, we saw the labor union secretary talking to the fired foreman. He asked her whether she knew about his transfer. She said that she had not known anything because she was on vacation. And now she had just returned. Then she said to the fired foreman: "You better go back to your work."

Later, the labor union secretary told us that she had been collecting signatures during the tram strike. The signatures were supposed to be attached to an appeal to the tramworkers to resume work. She went to the shop together with the departmental secretary. "We asked him to sign and he declined. And though the departmental party secretary stood at my side, he nevertheless refused to sign it!"

Once more we met the fired foreman; he was undecided as to what to do, wavering between rebellion and submission. He looked very tired and nervous, obviously under considerable mental stress. Referring to his refusal to sign the appeal to strikers, he said: "This was not my business. This was the business of those who worked in trams. Later, the section foreman called me to his office and they asked me whether I still insisted on refusal. I said, 'yes.' But, you know, the real reason for my removal is my October activities. We were a group that wanted to get rid of the plant director." He felt the futility of his fight against the plant director. He said: "The plant director reported me, saying that I was pro-American, that I was eager to get gift parcels from America. And doesn't it help the nation? Look at West Germany. They got American help and how well off they are today! Gomułka himself is doing the same today. Regardless of the social system, any help that gives something to the people is good."

Later, we went to see the new manager, together with the fired foreman. The new manager was surprisingly free to express his evaluation of the case. According to him, the real reason was that the fired foreman was too critical of the plant director. "As a matter of fact," he said to the fired foreman, "I understand your feelings. I was in a similar situation. But you cannot run your head against a wall. This would not help you."

At this moment I took the liberty to step in and say: "But this firing will make a bad impression upon the other foremen whom you try to win for cooperation."

The new manager: "You are wrong. The people must be afraid of some authority; they must fear the management." In addition, the new manager also explained why he did not come over to the informal meeting of the foremen in the foremen's dressing room. He said that there was no reason for him to come. "If I had gone, another group might organize another meeting, and so forth, and there would not be any end of such things. This I cannot permit."

The philosophy expressed by the new manager was clear and simple. It had probably proved itself to be a workable philosophy in many former situations. However, this was not the philosophy of the workers' council that called for discussions and consultations.

The case of the fired foreman was once more touched upon in our farewell call upon the plant director on September 3. The chief executive said: "He is a difficult case. One cannot talk to him, and he did not want to listen to us. Thus, we released him."

On this occasion, the director made a pessimistic statement similar to that expressed above by the new manager concerning the workers' cooperation in the management: "After October, the government hoped to get support of the workers, to raise production. And you can see for yourself what the results are."

From our own observation, we would conclude that the firing of the foreman was a loss to the cause of workers' participation in the management. The fired foreman was a highly articulate and obviously intelligent person who had some influence on his colleagues. In other words, he was a sort of informal leader within a clique of foremen. Foreman I praised him highly as "a courageous and honest man who was not afraid to speak out." Intimidated by his discharge, the remaining foremen were made even more silent. Or, in the opinion of I-5: "You see, the foremen are now afraid to speak out. But we workers, we are not. Only the workers can do something."

The major events reported in this chapter dealt with the great effort on the part of organized bodies, that is, the management, the workers' council, the party, and the labor union to get the workers more involved in the issue of production.

The initiative was steadily flowing down, from management to labor. And yet the case of collecting money for the section foreman showed that the workers could act on their own, though in a simple manner. In other words, they also could develop some initiative of their own. The problem is why did they fail to do so when attending production conferences. In the next chapter an attempt will be made to answer this and other related questions.

CHAPTER

5

Analysis of the 13 Events

CHAPTER III has sufficiently illustrated the point that the October change had actually brought no significant shifts in the personnel of management. The plant director survived the popular demand for his removal. And so did all around him. The removal of the manager of the weaving department (event No. 4) strengthened the control exercised by the director over the plant. Notice, for example, that in making reference to his intervention at the party meeting in behalf of the section foreman, I-5 referred only to what the plant director had said (event No. 7). Similarly, in discussing the removal of the fired foreman, the new manager spoke only about the plant director's decision (event No. 13).

Therefore, it seems reasonable to conclude that the October change had only subdued to some degree the plant director's dominant role in the factory. Comments made upon his strong vocal behavior at the foremen's meeting (event No. 2) showed clearly that persons were comparing the plant director's present behavior with his past performances, finding and pointing out similarities. The removal of the weaving department manager and the case of the fired foreman made persons aware that the plant director did not change radically his control of the enterprise in spite of the fact that there was definitely more freedom of expression; the past still lingered on.

During the "Stalin years" workers had developed certain habits that could not be changed overnight. Fundamentally, the attitude based upon former experiences could be described as one of distrust. Discussions held in the smoking room (for example, events Nos. 8 and 10) showed that our men felt frequently to be manipulated by "them." Former experience had shown to the workers that there was a great difference between the verbal definition of the situation as appearing in official pronouncements and the reality of their daily experiences. They sneeringly commented upon the official thesis that "workers rule the factory"; I-5 experienced concretely such discouraging disparity between the norm concerning the two-thirds proportional workers' representation in the workers' council and the actual composition of the presidium of the workers' council (event No. 9). While the poster concerning the illegality of obligatory Sunday overtime work was still hanging on the walls, the workers frequently experienced that this rule did not count in former days (event No. 8).

Hence, verbal pronouncements and promises were understood not to be taken at their face value but to be evaluated only in terms of actual events. No wonder, then, that the verbal appeals of the plant director and other leading figures were ineffective. The former experience of the workers had actually produced a devaluation of verbal pronouncements.[1]

Analyzing statements made during meetings, we see that the production meetings and the workers' council assembly meeting (events Nos. 1, 2, 12) covered many items, but there were very

---

[1] The impact of Gomułka's reappearance on the political scene was increased also by his open indictment of former official denials of facts; despite their debunking and cynical attitude, our men and women craved for ethical behavior.

few decisions, if any at all, resulting from these discussions. In terms of the extensity-intensity model (Figure No. 1, Introduction) discussions were more extensive than intensive. Also there was a prevalence of general exhortative statements as compared to descriptive-cognitive propositions. For example, undertaking an impressionistic content analysis of the workers' council assembly meeting (event No. 12), we are inclined to conclude that the following three categories can be ordered in terms of increasing frequency: (a) there were some technical statements on technical and financial production matters; (b) there were more frequent identification of responsibility talks; and (c) the most frequent were exhortative ethical appeals made to the audience.

If the three large group meetings (events Nos. 1, 2, 12) are analyzed in terms of direction of statements, we can observe that the verbal behavior flowed not only from the leaders of organizations to the rank and file, but also that most statements were, to a large extent, repetitious. Only in rare instances were there tendencies to comment on a prior statement; no real dialog and debate could be developed. Therefore, there was no increase in tension characteristic of the process of a group involved in the search for a solution.[2] The large body meetings were mostly composed of repetitious "monolog" statements, produced by managerial personnel. Being in higher leadership positions, all managerial persons probably felt obliged to make some statements when the rank and file kept silent.

Comparatively, the workers' council presidium meetings tended to be not only more intensive but also, to some degree, polemical (events Nos. 9 and 11). We saw that I-5 and the chairman of the weaving department workers' council were asking questions and also getting answers. Notice, however, that the actual decision was already made before the meeting

[2] See Robert F. Bales and Fred L. Strodtbeck, "Phases in Group Problem Solving," in Dorwin Cartwright and Alvin Zander (eds.), *Group Dynamics: Research and Theory* (Evanston, Row and Peterson, 1953), 386-400.

of the presidium. There was no vote taken despite the fact that the majority of the presidium's members probably favored the immediate distribution of the 13th wage (event No. 9). To some degree, the difference between the unfolding dialog of the presidium meetings and the pluralistic monolog of the large group meetings was due to the different size of the deliberating bodies. Notwithstanding the fact that the smaller presidium favored development of discussion more than the larger bodies did, the dialog of the presidium seemed to be arrested at a certain moment of its development.

Returning to our extensity-intensity participation model, it appears to us that even workers' council presidium meetings did not get beyond the first two degrees of the model, i.e., to be informed and to express opinions. The only exception was the issue of the reject tie-yarn boxes (event No. 11). Otherwise, during all production talks, one did not really get to the point of decision making.

Our analysis so far has shown that first, the management actually did not relinquish any significant part of its power and control of the plant, and secondly, that during the production meetings no decision was actually taken or shared with workers' delegates.

Considering the reasons for the lack of participation of workers in at least some decisions made in the plant, we should list three major factors: First, there was the organizational-institutional structure that legally made provisions for workers' participation while simultaneously functionally defeating that purpose. Secondly, there was the inertia and lack of interest and suspicions of the workers themselves that deprived them of any possibility to participate at least in some decisions. Thirdly, there was the personality of the plant director, who tended frequently to dominate the scene.

In weighing the importance of the factors that accounted for the unsuccessful verbal cooperation between management and workers, the organizational-institutional framework seemed to be the most important. The major reason for this unsuccessful cooperation was the ambiguous definition of the functions that the organizations had to fulfill. The ambiguity was derived from the Marxist-Leninist theory concerning the role of the party both within the state as well as within the industrial enterprise.[3] Accordingly, the labor union was ambiguously assigned to perform functions that are rather contradictory, e.g., protect the workers against management while simultaneously inducing them to comply with management's work demands.[4]

A good expression of the ambiguous goals was the plant director's statement concerning his preference for a foreman who does not fulfill the production plan but who participates in discussions (event No. 2). Although the major purpose of the production conferences was to increase productivity, there seemed to be also a demand for developing greater political loyalty among workers. From other analyses of the Communist policy, it becomes evident that political control sometimes goes ahead of economic expediency.[5]

The ambiguity of goals brought about a certain lack of delineation of functions that particular organizations had to perform. In Chapter III, the overlapping of functions of labor unions with those of the factory's social department was sufficiently demonstrated. The bylaws of the workers' council

[3] A good example of the multiple and overlapping criteria of successful management can be found in David Granick, *Management of the Industrial Firm in the USSR: A Study in Social Economic Planning* (New York, Columbia University Press, 1954), 284; Reinhard Bendix, *Work and Authority in Industry: Ideologies of Management in the Course of Industrialization* (New York, John Wiley, 1956), 413; Andrew G. Frank, "Goal Ambiguity and Conflicting Standards: An Approach to the Study of Organization," *Human Organization,* XVII (Winter, 1958-1959), 8-13.

[4] A similar ambiguity could be found for example in the official praise of the value of group discussion and action, and simultaneously in the *divide et impera* social control technique of the plant director (event No. 1), by which the natural solidarity of the work group would be undermined. Bendix, 431, points out that the party seeks to undermine the natural solidarity of the group in order to increase its control over individuals.

[5] Ivan Gadourek, *The Political Control of Czechoslovakia: A Study in Social Control of a Soviet Satellite State* (Leiden, H. E. Stenfert Kroese N. V., 1953), 198. See also, Thomas T. Hammond, *Lenin on Trade Unions and Revolution: 1893-1917* (New York, Columbia University Press, 1957), 126.

(Appendix, section 24) postulated that even the workers' council takes care of or considers certain social welfare functions normally reserved for labor unions.

It should cause no wonder, then, that there was a clear tendency to place responsibility for failures upon other organizations or persons (events Nos. 1, 2, 9, 11, 12). Although this ego-protective technique is a common human tendency, the functional diffuseness in the plant favored a frequent appearance of it. Surely, the plant director was the one who used it frequently. Conversely, he also served as a symbol upon which dissatisfactions of persons were projected. From the viewpoint of the workers, he was the man to be blamed. In all fairness to him, when criticizing his policies, people did not make allowance for the fact that the plant director was bound by existing institutions and orders coming from distant headquarters.

Generally speaking, we may propose that as the diffuseness and lack of visibility between the functions of an organization becomes greater, so will become the tendency to assign the blame upon a particular person or item instead of distributing it specifically to particular malfunctioning elements of the social system. The plant director served in this instance as a scapegoat.[6]

If this analysis is thus far correct, the following question could be raised. Why was there little conflict between organizations whose functions were overlapping? According to the Communist theory, the party was supposed to supply the integrative authority.[7] What was actually the relationship between the party and management in the plant?

We have noted that the plant director was the most often mentioned figure even when the party was talked about.

[6] Thus, our experience would lead us to qualify Margaret Mead's thesis about the Soviet diffuse notion of guilt. While Mead explained it as a result of former Russian cultural values, we would explain it as a property of the social system, i.e., its functional diffuseness. See Margaret Mead, *Soviet Attitudes Toward Authority: Interdisciplinary Approach to Problems of Soviet Character* (New York, McGraw-Hill, 1951), 27.

[7] Granick, 230.

Obviously, being a member of the party, he dominated the factory party organization. From the disrespect expressed toward the labor union by some workers and from our direct observations of the director's behavior in regard to labor union officials, we conclude that he also dominated the labor union organization to some extent. This concentration of power in the hands of one man and his more or less dominant influence in the three organizations had both functional as well as dysfunctional effects upon social processes in the factory. The functionality consisted in the fact that a possible source of conflict relationship between the party and management, described for example by Bendix, was eliminated in the plant.[8] The reader will remember that all three organizations, when facing the labor audience, maintained a united front. There was only one instance betraying a lack of solidarity between the management, party, and labor union personnel. During the poster incident (event No. 10), the plant director faced the angry crowd of workers alone, while the new weaving department manager, the labor union secretary, and the departmental party secretary stood aside. However, when expressing disapproval of the plant director's backfiring visual communication idea, the shift supervisor (event No. 10) put such disapproval in personal terms. Consequently, the disapproval was not expressed in terms of disagreements about boundaries of organizational competency, but in terms of undesirable personal properties.

Of course, our inability to attend party meetings deprived us of firsthand evidence for the assumption that there was no serious competition for power between the party and management. The only apparent struggle for power was exemplified by the plant director and the manager of the weaving department. But in this instance, both men were members of the party and both belonged to the top echelon of the management. Therefore, it was again a conflict between persons but not between organizations. The relationship between the

[8] Bendix, 431.

plant director and the plant party secretary seemed to be rather that of mutual cooperativeness with a slight dominance on the part of the plant director.[9]

The relationship between the party and management was diffused. The reader will remember that the departmental party secretary defined the role of the party as follows: "When something goes wrong, the party acts. Members of the party are supposed to have a higher responsibility." This is obviously a very diffuse relationship, because there might arise easily a disagreement concerning the definition of "wrong." Because of this and also because of the fact that management personnel were simultaneously members of the party, the behavioral solution was found in terms of personalities. Since the plant director seemed to have a more driving and aggressive type of personality, he dominated the scene. It is quite possible that in another factory, the relationship between the party and management could have been reversed.

The workers' council had some institutionalized provisions against the concentration of power in one person's hands. According to the bylaws (Appendix, section 6), the plant director was not eligible to become chairman of the presidium. On the other hand, he could have vetoed the decision of the workers' council by refusing to act upon it and eventually by appealing to the higher state authorities.

This last observation brings us back to the point about the dysfunctionality of the concentration of power, called in the Communist terminology the "democratic centralism."[10] We have observed, e.g., in event No. 1, that as soon as the plant director entered the scene, he began to dominate it with his verbal pronouncements, while the members of the weaving department workers' council kept silent. The same observation can be made concerning the meetings of the other two large bodies (events Nos. 2 and 12).

[9] See similar mutually protective informal agreements reported by former Soviet managers in Joseph S. Berliner, *Factory and Manager in the USSR* (Cambridge, Massachusetts, Harvard University Press, 1957), 266-67.

[10] Granick, 12.

Consequently, we propose that as the concentration of power in hands of the few becomes greater, so will become the tendency of the many to withdraw from participation or to express their dissatisfaction in sudden outburst of normless behavior. Both instances are exemplified in the workers' council large assembly meeting (event No. 12). There was the passive silence of the audience as well as the disturbing and unexpected invasion of noisy workers from the weaving shop.

Furthermore, it seems to be a functional property of the social system that a concentration of power, unless institutionally controlled, will tend to generate more power. We saw that during the workers' council presidium meetings, management sought to use the new organization in the same way in which it had been using labor unions, i.e., to communicate with and to control workers. On the other hand, I-5 sought to use the workers' council in the opposite direction, i.e., to express and implement desires and wishes of the crew (events Nos. 9 and 10). We also heard our men expressing their fears lest I-5 would go "over to management" (event No. 10). As a matter of fact, management tended to "coopt" persons that showed some abilities. I-5 was invited to join the party. If a person refused, the managerial party power center would eliminate him (event No. 13).

Thus, we are getting the somewhat paradoxical situation that the functional properties of the social system were counterweighing the legal norms and values of the workers' council. As a result of this "inherent contradiction," there was a permanent demand for the raising of morale. According to the plant director's statements (events Nos. 2 and 12), the purpose of the meetings was primarily that of morale building.

Regardless whether the management believed or not in the usefulness of the workers' council, the fact remains that all four organizations made a definite and outspoken effort to break through the barrier dividing workers from the managerial group. The leaders of the four organizations were continually initiating verbal messages directed at the rank and file. However, did the management realize that the effect of their verbal attempts might have been rather dysfunctional?

It seems from all evidence that management was aware of the negative attitude of workers. For example, the weaving department manager listed the negative attitude as one of the major obstacles to be overcome (event No. 1). The reader will remember that the second secretary of the party, at the conclusion of the workers' council assembly meeting, said: "It is necessary to explain where the money comes from. If we are right, why be afraid to go and see the people and talk to them?" (event No. 12).

It becomes possible now to attempt to explain why management personnel tended to use the same unsuccessful method of persuasion over and over again. Inasmuch as the Communist movement has developed a certain set of institutional behavior supported by the Marx-Leninist theory, a person is expected to live up to it in his behavior. "To mobilize the masses" is the common duty of the leader and of the members. Therefore, one can assume that several speeches delivered by lower echelon leaders were actually produced for the purpose of displaying loyalty, i.e., for the purpose of self-defense. If the result was economic failure, a person might feel still protected because he followed the theory's prescriptive pattern.

The fact that the social system is able to operate must then be explained in terms of the solution observed also in the plant. On the level of cultural norms, one pays lipservice to cooperation and the sharing of decisions, while in terms of the social system, one concentrates the power of decision in one center. In other words, there develops a great disparity between the symbolization of the actual state of affairs and its phenomenal counterparts.[11] No wonder then that this disparity

[11] Note that in our discussion of the history of Polish labor unions in Chapter I, it has been pointed out that the shop committee law was never formally repealed but made behaviorally invalid by new legislation. Furthermore, although our data cover mostly workers, we would be inclined to conclude that the management in the factory was bound by a great number of limiting and contradictory prescriptions that forced it to behave actually "illegally" in order to get results. See also Bendix, 384, and Berliner, 160-81.

has demoralizing effects, and that, consequently, the leadership feels also for this reason to be compelled to be permanently concerned with the morale attitudes of the workers.

Referring again to our review of the problem of workers' participation in management in the Introduction, we can conclude that our data give answers only to one aspect of the problem; i.e., that in order to help persons develop some degree of participation in decisions (and again we do not know this degree in terms of intensity-extensity), it is necessary to help them achieve a certain amount of identification with the production or task of the organization. Our analysis has brought to light the organizational and institutional variables that tended to reduce rather than to develop whatever potential propensity workers had to participate in management. There was a surplus of organizations that professed to cater to the needs of the workers; nevertheless, none of them was successful in the sense that it served as a group symbol for our men and women. The workers did not feel that they had really any organization of their own (event No. 5). Our earlier analysis of the Communist theory and institutions has shown that under the given organizational setup, this sense of belonging could not possibly be well achieved. Consequently, the answer was silence and organizational passivity.

Nevertheless, in one respect the official pronouncements seemed to have some small effect. Though workers sneered at such statements as "workers rule the factory," they used similar words sometimes to enhance their self-confidence when acting on their own. We have reported several times in the preceding chapter that I-5, almost an exception among the men of the smoking room, frequently emphasized and overstated that "only workers were not afraid" and that "only workers could do something" (events Nos. 10, 13). When invading the assembly hall during the workers' council meeting (event No. 12), workers justified their behavior by emphasizing that this was a "workers'" council. Thus, the management appeals had eventually obtained some results, though in an undesired and unanticipated way from the management's viewpoint.

The workers were also able to initiate action by themselves to some degree (events Nos. 6, 7, 12). While in event No. 7 our men and women used the management organizational structure to collect money in behalf of the section foreman, in events Nos. 6 and 12 they acted beyond and outside of the management structure. In event No. 6 they bypassed the management and went to plead in behalf of the section foreman directly to the party. In event No. 12 some of them invaded the assembly hall against the wishes of the management, the party, the labor union, and the workers' council. The invasion of the workers in event No. 12 echoed events that happened in the factory during the October upheaval (Chapter III). In their normless behavior, the workers acted outside the boundaries of the four organizations in existence at that time.

It should be, however, pointed out that the actions undertaken by the workers themselves were relatively quite simple and short termed. There was little planning, timing, organization, and consideration of particular aspects and consequences of their acts. To participate in problems of management, to sustain in them a genuine interest, and to be able to produce some fruitful ideas were probably beyond not only the level of motivation but also the abilities of some of the men and women. It is necessary to take into account the fact that physical laborers—and this holds probably true to different degrees all over the world—have on the average somewhat smaller ability to manipulate abstract concepts and to initiate new ideas. Laborers in this sense tend to be more conservative and passive (event No. 3).[12] In the self-critical words of the young man from group III: "The men are really not interested in management issues. They get involved only if their pockets are involved" (event No. 10).

Sociological characteristics of the labor force in the plant

_____

[12] Joseph A. Kahl, *The American Class Structure* (New York, Rinehart, 1957), 205-15.

also contributed to a relatively weak participation and interest. There were predominantly women and many persons fresh from farms. The average schooling for both groups was four and one-half years. Most of them were semiskilled, with a little of the craftsmanship pride in their jobs. Furthermore, their life experiences, consisting of oppressive years under the Nazis and Stalin, surely did not contribute to the development of a "free speech habit." And neither did the recent strike of the tram employees (event No. 5).

We have paid a great deal of attention to the role of the plant director in the preceding discussion. Such discussion became necessary because, within the structure of organizations, this role provided for a relatively considerable concentration of power. On the other hand, due to the fact that it was simultaneously controlled by other outside centers of power, this role was endowed with stress. According to the workers' council bylaws, the plant director's role was defined as "one person's directorship and one person's responsibility" (Appendix, section 30). Conversely, the issue of the leaking roof was considered to be beyond the plant director's decisional jurisdiction. It should not be strange, therefore, for the incumbent of such a role to have experienced strains and stresses of which other persons, especially the workers, were not quite aware.[13] Although the organizational-institutional framework ought to have been actually the primary target of criticism, the workers tended to blame the plant director.

Now, while it is true that the organizational setup of the plant director's role accounted to a large degree for his unpopularity with the workers, the second contributing factor was his personality. It is probable that the difference between the institutional role factors and the personality factors would appear in a more meaningful manner if we were to compare the plant director with the weaving department manager; we

---

[13] Characteristically, the Warsaw newspaper *Głos Pracy* counted 12,850 old regulations that could hamper activities of the workers' council. Quoted by Miedzińska, "Kryzys Samorządu Robotniczego," 75-76.

would find that both men were members of the party, that both belonged to the upper echelons of management, and that, nevertheless, the former was unpopular while the latter enjoyed certain popularity among the workers. Probably the reason for the plant director's unpopularity was that this short and stocky man tended to be somewhat authoritarian. People several times spoke about his "dictatorial practices" (event No. 10). We have noticed that the plant director tended to dominate the rostrum wherever and whenever he appeared. He spoke more and criticized more than he listened and praised. Possibly his relatively smaller verbal assertiveness during the presidium and assembly meetings (Nos. 11 and 12) was due to his bad experience with the poster during the previous day. He doubtlessly realized that the poster idea was a mistake and, on the next day, tried to be more conciliatory at both meetings.

In spite of the fact that the direct ethical appeals to the audience might have been a result of the "institutional behavior," on the other hand, because in some areas of social structure, such as the relationship between the party and management, there was an "institutional vacuum," the plant director also was compelled to develop a habit of self-assertiveness if he were to survive. If there had been prescriptions as to how to share power, etc., it is possible that the authoritarian personality tendencies of the plant director would have remained more subdued, winning him some acceptance on the part of the crew. Notice that the lower echelon persons sought to use different techniques when transmitting production demands to workers. The section foreman's personal appeal for Sunday overtime work (event No. 8) was an example of the opposite method to that used by the plant manager in his broad patriotic appeals.

It seems true, on the whole, that the plant director experienced some insecurities in his role because he obviously was eager to remove all his outspoken critics (events Nos. 4 and 13). On the other hand, he displayed personal courage when facing alone the angry workers during the poster incident (event No. 10). Furthermore, he withdrew the section foreman's release from the job (event No. 7). Under his leadership, the factory had achieved good production results several times. If we take into consideration that the plant director was responsible for a production unit with more than 8,000 persons who worked—according to American standards—with outdated machinery, and that he had simultaneously to deal with a great number of different controls, norms, and institutions and organizations that he had not designed and most likely neither desired, we get a more balanced view of the plant director's performance.

What are the results of our analysis of the 13 events? Suppose the plant director would have had a more pleasing and "democratic" personality; suppose that the workers would have been more educated and interested in management problems. Would then the workers' participation in management have significantly increased? I am inclined to maintain that unless there would be an institutional provision for the workers to have genuinely their "own" organization with which they would identify themselves and whose representatives they would trust, the participation would not be great. Of course, even if the latter would have been provided, it would be most likely a participation by means of representation only. But at least those among the rank and file who would have been motivated and would have had greater capacities would be drawn in.

6

Attitudes toward Production

WE HAVE SHOWN that the workers tended to perceive the function of the production process more in terms of personal interest rather than a collective goal. The workers did not identify themselves with the plant and its major function as did the management people. Since this was directly relevant to the problem of workers' participation in the management of the plant, we asked a series of three questions on this topic. The responses were not written down in the workers' presence, lest it influence their replies, but they were recorded immediately after parting from each respondent.

The first question asked was: "Whose factory was the plant?" The standard answer was: "It is the state's factory." Then I said: "Fine, but who is the state?" This question troubled my respondents a little. We arrived, however, after some clarification, to the point where my respondent agreed that every Pole, including himself, is part of "the state." Though such a conception of the state could be criticized from the legal viewpoint, it served well the purpose of asserting, "This means that the factory also belongs to you." As the responses to this statement provide essential evidence for the thesis that collective property does not necessarily result in the disappearance of the differentiation between management and labor, all answers will be presented here.

The six "social leaders" responded in this fashion:[1]

I-5: "This is state property. Of course, it is our property, but people do not feel like that. It will take many years before they change their attitude."

I-6: "The plant belongs to the state. The state governs. A house can be mine, but not a factory. If I stop work, I do not have any profit from the factory."

I-10: "Well, the state is everywhere. . . . The factory is my factory as long as I work here."

II-17: "What an idea! Only the plant director can have a feeling of ownership. This is his factory."

II-34: "Do you mean the people's state? Well, this means that the workers rule, but not the capitalists. Therefore, the factory is ours. But it is not mine. I do not know how to explain it otherwise."

II-32: "How come? My factory?" and she laughed.

The replies of II-32's colleagues in the "top producers" group were as follows:

I-7: "This factory is not mine; it belongs to the state. Of course, the state is composed of us, but nobody has anything. If the factory belonged to everybody, everybody would take a piece home and there would be nothing left."

I-8: "Well, I do not know how to say it. Of course, this is the state factory. It is ours providing we work here; otherwise, not."

[1] As stated in Chapter I, not all data collected and tables developed are presented to the reader because of their cumbersomeness and lack of direct relevancy. By means of direct observation and sociometric questionnaire and a questionnaire measuring the degree of knowledge about the factory and its institutions, we have established that I-5, I-6, and I-10 in Group I and II-17, II-34, and II-32 in Group II ranked at the top. We shall call these persons "social leaders." Those who had the best production scores, "top producers," were I-7, I-8, and I-2 in Group I and II-20, II-13, and II-32 in Group II.

I-2: "The factory mine? What a queer idea!"

II-20: "This belongs to the state. I have nothing but work here. I could also work somewhere else."

The remainder of the answers were recorded thus:

Foreman I: "The state, this is the nation. But I do not feel that the factory belongs to us. If they would distribute the profit among us, I might feel it."

I-3: "I feel that I have to work, but not that I own the factory."

I-4: "How can I know? Well, the factory is here and nobody can sell it. . . . In the village, there was one fellow who drank up all his property."

I-14: "The state, this is all of us, we Poles! But here I cannot even take a small piece of textile with me."

I-22: "How could I be the owner of the factory?"

I-16: "This belongs to the state. How could it be mine? Mine is only that which I earn; otherwise, nothing."

II-15: "The state is all of us. But the factory is not ours. If I could earn more, possibly I could have such a feeling. The plant director can have such an idea."

II-11: "It is not ours; this belongs to the state. If I take a small piece of textile, they arrest me. If the director takes a full car of textile home with him, nothing will happen to him. Absolutely, this is not mine. Were it ours, they would give us different treatment."

II-24: "Oh, not at all mine. But if we did not work, the factory would stand still. It belongs to the director! He takes out money from the bank."

II-9: "This is state property. But it is not mine. I cannot take even a tie yarn."

II-26: "This is a common good. This is how one says it. This is not mine, but a common property."

II-22: "This is not mine. The only thing that I have here is my work, and they pay me for that, and that is all."

In reviewing the statements, one can conclude first that, with the exception of I-5 and II-34, there was nothing that would show that the respondents conceived of the factory as "their property." When they said that the factory was theirs while they worked there, they referred to the locale and not to the ownership. That the factory was indirectly owned by them appeared almost incomprehensible to some of them.

Exceptional is the answer of II-34, who was a young unmarried girl, educated in the home for orphans. When she made the statement, she appeared to be repeating something she had learned in the school, without any strong conviction.[2]

Comparison of the six social leaders with the six social producers does not disclose any significant difference with the exception of I-5, the delegate, and II-34.

Thus both groups showed an absence of identification with the factory. The reader might remember, however, that the men were bitter about the wasted money on badly repaired roofs and on real or imagined mismanagement.[3] On the basis of such statements, one could hypothesize that there was at least some potential identification with the factory. One should notice that several men stated that they might have such a feeling if they could earn more. The major idea of the new workers' council was exactly to develop such an identification through sharing in profits. In the plant the effect of sharing was, however, not yet felt; therefore, there was almost a complete absence of ownership identification.

For anybody working in the plant, this was naturally nothing new. From what we have said in the Introduction, it would appear quite natural that workers within an organization of more than eight thousand persons, with little participation in decision making and experiencing many human relations mistakes on the part of the management, would not be willing to develop the desired identification. By this we do not maintain that such an identification could not be developed. But a prerequisite for it would be a proper communication between

[2] Hypothetically, I explain it as a result of her sudden fears concerning my true identity.

[3] It should be, of course, recorded that the interest in the factory appeared only when the 13th wage issue came up.

workers and management. And this was, as was shown in several prior instances, absent in the plant to a considerable degree.

At this point it is interesting to see if there was a correlation between the degree of identification with the factory and the amount of production by an individual worker. Figure VI shows the order of persons in both groups according to their average percentages of fulfillment of the production plan over a period of eight months (January—August, 1957).

FIGURE VI

AVERAGE FULFILLMENT OF THE PRODUCTION PLAN BY
PERSONS IN BOTH WORK GROUPS IN PERCENTAGES

Group I			Group II	
Person	Percentage		Person	Percentage
I-7	107.3		II-20	106.8
I-8	106.0		II-13	106.5
I-2	103.0		II-32	106.3
I-6	101.3		II-17	104.2
I-4	100.6		II-15	104.2
I-5	98.2		II-9	101.7
I-10	97.5		II-26	99.5
I-14	96.1		II-11	97.1
I-3	91.1		II-34	96.3
I-16	86.3		II-24	95.8
I-12	83.3		II-30	93.0
I-18	83.2		II-22	91.2

The motivations of the three top producers in each group were sought. I-7, who had a family of a nonemployed wife and two children, perhaps naturally felt a great urge to work. He was sure in his opinions, being simultaneously blatant and uninformed. I-5 held a fatherly, critical attitude toward him: "Oh, this I-7, he works like a horse but has no idea about anything." When asked whether he was purchasing newspapers, I-7 answered: "No, I could not buy other items."

I-8, a recently married woman who had moved to Lodz a few years before, was doubly valuable, for she urged her husband, II-17, to work hard also. She did not identify herself with the plant; she still was "a villager in the factory."

I-2, the mother of two infant children, suffered frequently from a lack of sleep because the children kept her awake. She did not know much about the factory and did not care to learn, being absorbed by her family problems.

II-20 was a 22-year-old single woman of German descent from the former East Prussian territory. "You know," she explained, "I have to send money to my family; my father was killed in war." She did not understand and was not concerned with social and political issues.

II-13, father of a child and a former member of the police, had the most formal education. He felt that he could not talk to anyone in the shop because—in his opinion—the people were uneducated and did not know anything. However, he himself did not know anything either, as it was disclosed by the questionnaire measuring the degree of knowledge about the factory. He kept to himself at work.

II-32 was a young, unmarried girl, commuting from a little town close to Lodz. However, she was also among "the social leaders" in her group, participating in clique behavior and displaying relatively a high degree of information. Compared to her, all other top producers in both groups were ranked low in terms of social participation and degree of knowledge about the plant.

The inspection of Figure VI also shows that those persons whom we had classified as "social leaders" (I-5, I-6, and I-10, and II-17, II-34, and II-32) kept themselves in both groups (with the exception of II-34) rather above the median position. Obviously, the social leaders were not best producers, but neither were they the worst.

The conclusion we can reach from the production data is rather puzzling. The persons who were likely not to be motivated or be able to participate in some management decisions could, nevertheless, be top producers as far as the physical output of work was concerned.

When the absence of identification with the factory was discovered, a further question almost spontaneously presented itself. How did the workers feel about their work? Did they see its social usefulness? Did they feel a part of a production collective?

While interviewing the persons, their answers were this time immediately recorded in their presence. The questions asked were: "How do you like your job here? Why do you work?" And these are again all answers, because of their relevancy to the thesis about the difference between management and labor. And again, six top social leaders will be presented ahead of the six top producers and the other persons in both groups.

I-5: "Well, I want to produce as much as I can. That's all."

I-6: "Look at I-2. She is twenty-nine and a wreck. How will she look in her forties? This job here in the weaving department is not very gratifying. Broken threads make my nerves tremble. If I could work in an office, I would not be this tired. When I get home, I do not know where to put my hands. All of us are tired. Before the war, I was eager to work. Now, I have to work."

I-10: "I have to work and have to be satisfied because I do not know any other trade. If one doesn't work, one doesn't get anything."

II-17: "I am not satisfied with this job. I do not get what I need. At home, I have to care for the child, prepare breakfast and lunch for it. Thus, I am already tired when I go to work. I would like to become a foreman."

II-34: "I have been trained to be a weaver. I must work. If I do not work, I cannot stay in the workers' hotel. When I was young, I did not have any choice; I had to become a weaver."

II-32 (also a top producer): "Now I am satisfied because Foreman II is good. He knows how to repair the looms. I do not earn bad money at present."

I-7: "I work because I want my children to have something to eat. I want my wife to stay at home with them. If the wife doesn't stay at home, the children get into trouble. I like the job and have the desire to do it. I would like to become a foreman. Possibly, I shall become one sooner or later."

I-8: "I like the job. I do not have any complaints now. My neighbors envy me because I produce more than they do."

I-2: "I don't know whether I like this job or not. I have to work because of my children. My husband doesn't earn enough. It is bad at night because I do not get enough sleep."

II-20: "I like to work. You know that I am a good worker, that I can achieve it."

II-13: "The job is not interesting. But it is necessary to work. I do not have patience to bind the broken threads. This is a job for women. I would like to quit and go into construction work where one works under the open sky. Of course, it is a heavier job, but one earns more. I am not afraid."

I-12: "I don't know why I work. My eyes are bad; my feet are weak. I work in order to be able to live. This is a tiresome job."

I-3: "To work, this is a necessity. But I am not interested in this job. I would like to become a foreman. I have been a weaver for fifteen years; that is enough. I am fed up with this work. It is always the same. The foreman has a little variety. Once that, another time something else. But this is only work, work, and work."

I-4: "I work because I must work. I would like a lighter job. I would like to work at home. This is a heavy job."

I-14: "I want to work as much as possible to earn more. I shall not change this job. In every job it is necessary to work."

I-16: "The work is not pleasant here. The steam goes on the wall; it is warm and dry. The threads break today in a fatal way. But, I must work."

II-9: "This is an uninteresting job. I work in order to live and not die of hunger. But I would like to get a supervisory job. But they did not want to release me."

II-11: "I work well. I have been working for ten years. This is my profession and I have been practicing it for fifteen years. I could not take any other job. I am interested in this work. It is necessary to work. Man must work."

II-26: "I don't know what to say. If I do my laundry on Monday and we have a night shift, I feel very tired during the night work."

II-15: "I must like this job. There is no way out. Where should I go otherwise? If I were working in a wool factory, I might earn more. A friend earns some sixteen hundred zlotys. But it is difficult to get such a better paying job."

II-24: "I must like this job. I would like to have a cleaner job. Here I have to move so much that I get tired."

II-22: "I like this job. The trouble lies in the bad warps. It is better to be a helper, but now I am earning two hundred zlotys more, and that counts. Of course, it is a tiresome job. I do not have time to think of anything else other than the job."

The conclusion that can be made from the above data points to a resignation on the part of the workers. The work is a "must." There is no escape from it. The complaint of monotony appears in several responses. On the other hand, those respondents who said that they liked the job did not specify what they liked about it. It is probable that a further probing on the part of the questioner, would have shown that they were not enjoying the job as such, but some other aspects, such as security and the like. Note that I-8 liked the job because she felt herself to be a better producer than other persons. II-22 liked the job because, compared to her former pay as a helper, she was earning more. II-32 liked the job because she felt that her foreman was better than other foremen. These preferences are not related to work as such.

The perception of work was defined strictly in individual terms. If we remember how much the plant director emphasized the collective goal of production, the absence of such a collective definition on the part of the workers is the more striking.

It can be also readily seen that there were very few references to personal relationships. Only I-8 and II-32 mentioned other persons in the group. All other respondents referred either to the inescapable necessity to work, lack of choice of other jobs, low or better pay. Characteristically, only men showed some aspirations to move to other occupations, to advance within the factory.

Comparison of social leaders with top producers discloses that more social leaders were dissatisfied with their work than were top producers. As the sample of social leaders and top producers is extremely limited, the difference between both subgroups can be considered no more than suggestive.

The finding concerning the strictly personal definition of the function of work and the absence of personal relationships during work in the shop led us to the hypothesis that the work group as such would not be perceived as a force or cause accounting for troubles or success in production.[4] Consequently, we started to inquire into the workers' explanation of their production troubles.

There were four major questions asked by us in structured interviews. First, showing the workers their fulfillment of the plan over the last six months, we wanted to know what their reasons were for failing to meet the plan during some of the months. Secondly, we went over with them the daily percentages of the realized plan from August 1 through August 22. Finally, we asked on each of the days from August 26 through August 31 how they had worked the day before.

When surveying the months of January through June in terms of achieved percentages of the plan with particular workers, it was surprising to find that they did not know, in most cases, how to explain the variation in their production achievement. In Group I, seven persons did not have any explanation, while four persons listed their own absences as causes which contributed to the failure of the plan. I-2 listed nursing her infant; I-8, sickness; I-5, attendance at workers' council meetings; and I-7, simply absence.

In Group II, again four persons gave some explanation, while seven persons had nothing to say. The reasons were similarly

---

[4] Another index of the absence of personal relationships as a part of the work process was the absence of helping each other. We observed directly that there were only three instances of mutual help, though there could have been, theoretically, 39 possible instances.

personal, with the exception of II-13, who listed for May the deficiency of harnesses. Sickness was listed by both II-11 and II-17. II-9 listed his marriage as the cause of his drop in production.

The workers could remember no more efficiently the causes of their difficulties during the last four weeks.

FIGURE VII

CAUSES OF LOW PRODUCTION ON PARTICULAR DAYS
OVER FOUR PAST WEEKS

Causes	Times Reported
Failure of the warp	21
"I do not know"	18
Personal	11
Repair of the loom	10
Down time	2
Broken threads	1
"It went badly"	1
Trapped shuttle	1

The "repair of the loom" category should be discounted; it was listed eight times by I-3, who used it most likely as a stereotyped answer without any great consideration. Under "Personal" were listed absences or early departure to take a train, but never indisposition during the work period, lack of concentration on work, and so forth. In other words, no subjective, self-critical causes of poor production were listed. Note that also there had not been any reference to personal relationships within the group. Failure of the warp was deferring the responsibility to another unit within the weaving shop.

Since it appeared during the interviews that the respondents could not remember efficiently the causes of their failures, frequently not even if the time was just two or three days ago, we systematically interviewed both groups each day from August 26 through August 31. Thus, responses for six days were obtained, that is, from Monday through Saturday. We asked: "How did you work yesterday?" "What were the reasons for any trouble?" "Did you meet the production plan?"

The respondents evaluated their work in different terms. We have reduced the answers to positive, neutral, or negative evaluations. Figure VIII shows the results.[5] The data showed

FIGURE VIII

PERCEPTION OF WORK PERFORMED THE PREVIOUS
DAY, REPORTED FOR SIX DAYS

Days	Good	Undecided	Bad
Monday	10	5	5
Tuesday	11	7	4
Wednesday	11	3	7
Thursday	3	3	15
Friday	7	3	7
Saturday	9	5	6

an absence of any definite pattern. Only on Thursday did most persons report bad work performance. Analyzing the data from the viewpoint of individual persons, no relationship has been discovered. For example, I-7, the best producer in Group I, had three positive and three negative evaluations. Of the two worst producers, I-14 had two positive, two undecided, and two negative evaluations; I-12, one positive and five negative evaluations. Thus, there seems to be a pattern of irregularity in the perception of work performance.

Seven categories were listed spontaneously by the respondents when asked the reasons for a poor performance.

FIGURE IX

REASONS FOR BAD WORK PERFORMANCE THE PREVIOUS
DAY, REPORTED FOR SIX DAYS

Reasons	Times Reported
Broken threads	33
Fixing of new warp; or bad warp	14
Loom repair	12
Trapped shuttle	8
Temperature, good or bad	6
Lack of interest because of unpaid 13th wage	3
Foreman unable or slow	3

[5] Only on Tuesday were there no absentees. Therefore, the total replies for other days was not 22, as it ought to have been for both groups at that time.

The broken threads are clearly the most important cause as seen by most persons during the week under study. Except for the three references to the unpaid 13th wage, no mention of psychological or human relations problems was listed. This is a picture observed earlier. The respondents explain production troubles only in technological categories, but seldom in social relations categories. Furthermore, additional data were obtained showing that the workers' achievement depended upon random factors which were not controlled either by workers or by management. Technological failures could have occurred any time.

The randomness was, of course, a contradiction to the desired control of work as sought by the plan. How did the workers perceive the plan?

It should be noted that since October, 1956, the plan's fulfillment had become irrelevant for the amount of payment. Workers were paid on a piece-rate basis, whether they met the plan or not. Therefore, it is not surprising that when we first asked whether they met the plan or not, six persons in Group I and five persons in Group II did not even know what the plan was. As we continued asking the next days, they started to pay greater attention and reported their achievements in terms of the plan. In a total of 21 times out of 66, persons in Group I did not know the plan or did not mention it. In Group II, persons failed to know or report the plan 14 times out of 66.

A follow-up question was directed to the acceptance and popularity of the plan. The answers again were classified into three categories: 5 positive, 11 neutral, and 6 negative. Most of the persons worked as much as they could anyhow; thus, the plan was useless from their viewpoint. It was another institution that surrounded the worker and that he felt to be superfluous.[6]

The procedure to determine the difference between man-

[6] Another index of the "oversupply" of organizations was the failure of the workers to differentiate between the labor union and the workers' council. Asking for the difference in the "degree of information questionnaire," we got the "correct" answer only from Foreman I and I-5.

agement and labor on the perception of production troubles in
the weaving shop was relatively simple. Twenty causes derived
from pretesting were recorded on cards. (The twentieth was
the open-end "other cause.") Through pretesting we also
discovered that respondents found it easier to rank several
causes at the same level. Therefore, we asked a respondent
to select two major causes of production difficulties. Then
we asked him to choose three more additional causes which
could stand in the second place according to importance.
That our list was exhaustive is shown by the fact that only
one person selected "other causes"; this, according to the
section foreman, was poor training of the weavers.

FIGURE X

CAUSES OF PRODUCTION TROUBLES AS RANKED BY
MANAGEMENT AND AS RANKED BY WORKERS

Cause	Manage-ment	Workers	Six Top Pro-ducers	Six Top Social Leaders
1. Bad filling	0	8	4	0
2. Bad warp	1	14	6	0
3. Improper temperature	8	23	4	5
4. Old model looms	3	8	2	1
5. Lack of cooperation between shifts	6	3	2	0
6. Unwillingness to ask the foreman for help frequently	0	6	0	1
7. Excessive waiting for the helper	0	3	1	1
8. Excessive waiting for the foreman	2	6	2	0
9. No help from neighbors	2	1	0	0
10. Three-shift work	14	21	0	7
11. Lack of sufficient number of workers	22	20	5	3
12. Own farm work	4	0	0	0
13. Own home work	2	5	0	2
14. Bad apartments	6	15	6	4
15. Commuting from beyond Lodz	2	4	2	1
16. Lack of work motivation because of low pay	2	18	2	8
17. Part-time jobs	0	1	0	0
18. Frequent sickness	12	8	1	1
19. Bad quality of yarn	4	25	5	9
20. Other causes	2	0	0	0

Figure X presents scores for four different groups. First, there is the group of fifteen management people, starting with the plant director and ending with the section foreman. The reader will remember that the workers perceived the management, that is, "them," as beginning with the section foreman. Next are the workers from Group I and Group II, including their foremen and helpers. Finally, there are the six top producers—three from each work group—and six "social leaders" —again three from each work group.

The scores were weighted. The first two choices counted as two each, and the second three choices as one each.

Figure X becomes more meaningful if we arrange the causes according to the rank. For each group, three highest causes are listed in Figure XI.

FIGURE XI

THREE MOST IMPORTANT CAUSES OF PRODUCTION TROUBLES
RANKED BY MANAGEMENT AND WORKERS

Management	Workers
1. No. 11, lack of sufficient number of workers.	1. No. 19, bad yarn.
2. No. 10, three-shift work.	2. No. 3, improper temperature.
3. No. 18, frequent sickness.	3. No. 10, three-shift work.

Six Top Producers	Six Social Leaders
1. Nos. 2 and 14, bad warp and bad apartments.	1. No. 19, bad yarn.
2. Nos. 19 and 11, bad yarn and lack of a sufficient number of workers.	2. No. 16, lack of work motivation because of low pay.
3. Nos. 1 and 3, bad filling and improper temperature.	3. No. 10, three-shift work.

Inspection of Figure XI shows some significant differences between the groups. First, there is a clear difference between the management personnel and the workers. The management persons listed broad social factors as causes, but no technological factors, as the workers did.

Secondly, all three groups of workers mentioned bad yarn

as one of the major causes. Furthermore, the workers tended to list more technological causes, and less social-organizational causes. The reader will remember that the same was found even in a more pronounced manner in the detailed analysis of the causes of low production (see Figure VII).

Thirdly, there is a tendency to list causes for which other groups might be called to be responsible. Note that the management people listed sickness, that is, the frequent absenteeism of workers, as the cause of troubles. On the other hand, the same cause was listed by the workers as a rather less significant reason. Contrariwise, the workers listed the technological failures very frequently, while the management considered them as less important. The reader will remember that on several occasions the plant director maintained that the technological factors were under control. The workers did not accept this point and continued on several occasions to raise complaints of poor filling, warps, yarn, and so forth.

Fourthly, although there are differences between the management and the workers, they are agreed on numbers 11 and 10. The lack of sufficient workers was a factor that the management could not easily get under control because of the general lack of labor supply in the labor market at that time in Poland. The swing shift, however, could probably have been replaced by a relatively more permanent shift. The greatest problem was getting used to the night shift. As a matter of fact, it was hardly possible to do so within a mere six days of night work. Thus, persons working during the night shift were struggling more or less with the problem of lack of sleep.

Fifthly, the management people were more definite and agreed on the causes of the troubles than the workers. The workers were more distributed in their opinions. Only the six social leaders were more agreed.[7]

[7] During our introduction to the workers (see Chapter II), they told us spontaneously their complaints, identified as a technological theme, a low-pay theme, and a disrespect-for-the-management theme. To a certain degree, these are also the three major causes of the production failure as listed by the six top social leaders.

Sixthly, the comparison of the six top producers with the six social leaders discloses that the producers were more concerned with technological causes than the social leaders. This difference corresponds to our former findings that the top producers were less interested in their social environment and more concentrated upon work as such. The reader will remember that more top producers tended to express also a greater satisfaction with their work than the social leaders. As the social leaders tended to be more exposed to messages sent to the workers by the management than the top producers were, an almost paradoxical conclusion seemingly appears: Those who are more stimulated by messages and who participate more in discussions produce less than those who isolate themselves from such an influence.[8] Such a conclusion is, however, untenable, because not all persons who tended to be more isolated from the social influence of the group were top producers, but in several instances were the worse producers (I-18, II-30). Conversely, the two social leaders in Group II (II-32, II-17) had relatively quite high work achievement. Thus, our data allow us to conclude nothing but that under certain conditions a top producer must not be necessarily a person who knows much about the work group and organization of work and who participates in interaction with other members of the work team.

More securely we can conclude, however, that the data presented here substantiate the thesis concerning the difference in attitudes between management on the one side and workers on the other. The common ownership has not eliminated antagonistic relationship between both major groups of the studied industrial organization. Both groups perceived production problems in significantly different ways. Char-

---

[8] Of course, there seem to be more factors involved in the tendency of the social leaders to be less productive and less satisfied. A person who tends to be more involved in conceptual and symbolic behavior—and social participation and decision making are made up of such behavior—might dislike more the mechanical work. I-5 would be a good example. This problem is, however, beyond the scope of this study.

acteristic of the workers' perception was a lack of any collective orientation. They looked at work in individualistic terms, refraining from defining its social purpose and usefulness, stressing rather its chanceful nature while failing to remember its dynamics. Contrariwise, the management emphasized the collective nature of the work, its ordered planning and social purpose.

7

Interpretation and Conclusion

IN THIS STUDY two kinds of information were gathered. First, there were data on actual events which occurred in the factory over a period of almost one year. Some of the events were reported; some were observed directly by the researchers. Second, there were questionnaires and production data, most of which cover attitudes toward certain issues within the factory. In other words, we have the behavioral data, supplied by events, and attitudinal data, provided by the schedules and questionnaires. Fortunately, both the attitudinal and the behavioral data point in the same direction. There is no contradiction between them; this surely increases the validity of their evidence.

The general conclusion concerning the workers' council in the factory is that it did not function well. Of course, the workers' council was a new organization, just born, struggling with the negative heritage heaped up "under Stalin" in the hearts of the workers. On the other hand, the management personnel themselves probably were not quite convinced of the usefulness of the new organization. They also failed to exploit the better human relations opportunities provided by the new organization, despite their manifest effort to do so. Nevertheless, a social scientist cannot be satisfied with such a general negative conclusion. He must ask himself what were the greatest obstacles in the smooth functioning of the council? What could be improved?

In Chapter I we have asked four broad questions to which this research has been directed. Here are the answers:

Question I: We have found a dichotomy in the outlook on the major function of the whole factory. While management tended to define the function in terms of the collective goal of production, labor looked upon the same from an individualistic viewpoint. The workers worked in order to earn a living and hardly "to increase production in order to supply enough goods to the market."

Question II: If we ask for the cause of this difference in perception of the respective roles, it is necessary first to stress that the fact that the factory was state property, i.e., the property of all, did not abolish the differentiation between the managerial group on the one hand and the labor group on the other. Workers still felt that "they" were trying to take advantage of them, depriving them of a part of their deserved earnings. Therefore, it is legitimate to conclude that if the means of production becomes collective property, this does not necesarily result in an individual person's identification with the plant. The "ownership by all" is too diffuse to constitute a stimulus to develop individual ownership identification with the common property. By this is not meant that such an ownership identification could not be developed. But our data point out that there must be other factors present in order that such an identification could come into being. Possibly, collective ownership is not even the major factor for development of such an attitude of identification which would result in a feeling of responsibility on the part of the workers toward the factory.

Our case study hints that probably a more important factor is communication. In the plant there were barriers between labor and management. These barriers were constituted by the workers' distrust toward management and workers' lack of motivation to acquire necessary information. On the other hand, the management continued to use methods of persuasion which were ineffective. Despite the fact that the management asked the workers to share in making decisions, there were no decisions actually made with the workers' participation. The disparity between the verbal definition of the situation and the actual situation contributed to a deflation of messages sent by the management to the workers.

While in the West, labor frequently struggles to get access to management and its decisions and information resources, in the plant during our period of study the situation seemed to be reversed. Management, supported by three other organizations, sought to establish communication with the laborers, who were reluctant to respond. Initiation of action went mostly from the management down to the laborer. A characteristic of the factory situation was the fact that the impulse to form a workers' council came from the management and the party. Since the party to a large degree overlapped management in the plant, the fact remains that the workers' council, which was supposed to represent labor first, was organized by management and the party. In the United States there have also been attempts made by management to organize their own labor unions. The workers did not develop the necessary identification with such unions, now illegal.

Question III: To what degree has the organizational-institutional structure contributed to the unsatisfactory labor-management relations? Our case study shows what we have called the "oversupply of organization."[1] We have noted that the roles of the labor union and of the workers' council were somewhat overlapping. In the minds of the workers, there

[1] The same was found by Jan Szczepański in the field of administration, public welfare, cultural and social life. See his "Próba Diagnozy," *Przegląd Kulturalny,* VI (September 5, 1957), 6-7, 11.

was a great deal of duplication and confusion. Of course, the workers' council was a new, still experimental organization. During the assembly meetings, however, or the smaller production meeting, representatives of all four organizations concerned were addressing the workers in the same way. Somehow, from the workers' viewpoint, management, the party, the labor union, and increasingly the workers' council were speaking the same language; they appeared to be parallel bodies.[2] Consequently, the workers did not have any organization strictly of their own.

Thus, we get the paradoxical situation that there is a superfluity of organizations that profess and try to cater to the workers' needs, and yet none of them actually does so. Though management emphasized group discussion and group consultation, the basically collective nature of all work, it did not succeed in harnessing the power of the group for increased production. Some persons who actually participated least in group behavior were the top producers. Obviously, the restrictive or the stimulative power of the informal group, as found in American sociological studies, was more or less absent in the groups studied. Work was an individual business in which one struggled with the hazards of unpredictable technological failures.

Outstanding was the role of the workers' council delegate, I-5. He sought to break through the communication barriers between both groups, occasionally experiencing frustration in being misunderstood. So far as he was concerned, however, there appeared another problem for the theory of workers'

[2] This was legalized in 1958 by the creation of the Conference of Workers' Self-Management. This body is the highest organization in the whole plant, as it is composed of all members of the workers' council, of the party organization, and of the labor union. According to this new arrangement, the labor union ought to be concerned with the problems of production as well as with the defense of workers' interests in relation to the state. See W. Gomułka's outline of the new organization given at the conference of the labor unions in April, 1958. Warsaw Trybuna Ludu, April 15, 1958. A good critical analysis of the new economic institution can be found in J. Miedzińska, "Likvidacja Samorządu Robotniczego," Kultura, No. 6/128 (June, 1958), 105-13. It is probable that the new organization means in fact that the role of the workers' council has again been curtailed, as in 1947-1950.

participation in management. Though he actually took part in several meetings and was really involved and interested in what was going on in the factory, his productive achievement was only a little above the average. This raises another problem, the relationship between conceptual and symbolic behavior on one hand and physical work on the other hand. This problem is, however, beyond the scope of our study.

In general, we have seen that the organizational structure in the factory not only permitted but also functionally promoted concentration of power in the hands of the plant director. Also, the personality of the plant director tended to accumulate rather than to share power; the result was the dysfunctional effect upon the initiative of the workers and other management personnel. We have shown that the new organization of the workers' council was wavering between becoming another "department" of management or a real "workers'" workers' council. Management in its drive for more control was actually simultaneously defeating its attempt to win the workers over to a greater verbal participation during meetings. Both goals, domination and cooperation, are incompatible. By seeking to achieve both, management's actions were ambiguous, resulting in unanticipated consequences.

It should be recalled, however, that the management was itself also controlled by high and distant centers of power. The plant director's role was full of ambiguous demands. He was responsible to both higher authorities and formally to the workers' council for economic results of the enterprise, and yet he could not, for example, decide whether a leaking roof should be immediately repaired or not. In their criticism of the plant director, the workers did not realize fully that his behavior was more determined by institutional and organizational variables than by his personality.

On the whole, our experience has led us to conclude that organizations were more important and relevant for the problem of production than small groups. The reader will remember that we started the research with a close concentration on small groups, but that gradually we moved to the study of organizations, especially to the analysis of functions performed by the workers' council.

To account for all difficulties only because of the ambiguity of the organizational framework would be inadequate. In our analysis in Chapter V we have found that not only the personality of the plant director but also the inertia of the workers contributed to the failure of the workers' council. In several instances, workers' inertia and lack of interest in management affairs excluded a priori even the smallest degree of participation. Though the workers were able to develop some actions of their own that characteristically tended to run outside or beyond the existing organizations, these actions were rather simple and short-termed. To participate genuinely in management, to sustain in analytical conceptual analysis, was beyond the motivational and probably also the ability levels of some of our men and women.

Thus, reflecting summarily upon the institutional framework of the workers' council and the three other organizations, we are led to conclude that it not only permitted ambiguous definition of roles and goals, but that it also based its theory more upon desiderata than upon actual potentialities of the workers. In terms of our extensity-intensity model of sharing of decisions in industry, the workers did not reach even the two elementary levels, i.e., to be informed and to produce opinions. Obviously without these two prerequisites, no participation and sharing in decisions is possible.

Question IV: Considering the last question, what improvement could be suggested as a result of our study, we propose that management probably could have improved personnel and social organization policies even though the organizational-institutional dimension was outside its reach.[3]

[3] A similar limitation was emphasized by several critiques of the failing role of the workers' councils. See, e.g., Piotrkowski, "Samorząd Robotniczy a Zarządzanie Przedsiębiorstwem," 35-48. The author finds that the workers' council shows little activity because the enterprise (1) cannot get more raw material; (2) cannot fire superfluous employees; (3) cannot choose the kind of product to be produced; (4) cannot decide on the price of the product; (5) cannot determine wages and salaries. The enterprise can only (1) save raw material; (2) start side-line production.

There seemed to be a tendency on the part of management to remove persons who were classified as "troublemakers." But the latter were usually informal leaders, or potential informal leaders. The case of the fired foreman serves as an example of this point. Since during the "Stalin years" removal was the common managerial practice, the effect could have been the same: to silence the laborers.

It seems that management did not understand well (or was institutionally bound) that people do not change their attitudes and habits overnight. Despite the "Polish October Revolution," labor did not forget its former treatment. Though the plant director might have raised his voice because of his personal enthusiasm when speaking at meetings, the workers would interpret it "as a return to his former dictatorial way of management." From the viewpoint of human relations strategy, anything that reminded a person of the former Stalinist practices should have been carefully avoided. Therefore, the hasty election of the foremen's representatives was a mistake. Likewise, the omission of a vote on the issue of the 13th wage was also an error.

Furthermore, an assembly and public appeal was not the appropriate method in a situation characterized by great barriers standing between management and labor. Possibly, an informal approach would have brought better results. We observed that the representatives of the four organizations, with the exception of the section foreman, did not care to go down frequently to the looms and talk to people. The labor union secretary was infrequently seen in the shop during our stay there. Her office was in another building, and we understood that she was very busy with officework. The party secretary spent most of his time in his office, too. I-5, the workers' council delegate, communicated often with "the smoking room group" of men, but he failed to cover other men and women in the ten work groups which he represented. On the other hand, it could be illustrated by the section foreman's appeal for Sunday overtime work that the personal approach could make a difference.

Another point concerning the initiation of action by labor is related to the concreteness of suggestions and their implementation. There were several points frequently raised at the meetings, some of them more general, some concrete. The discussion of these points, however, frequently clouded in ethical and patriotic appeals, was probably dysfunctional. People mostly endured them passively. Furthermore, even if some concrete suggestions were accepted by management, no promise of implementation was made, or the implementation was promised in general terms and easily forgotten. The only exception was the immediate installation of the reject tie-yarn boxes, which was itself an earlier suggestion which management had left unimplemented for a long time.

We could advance a generalization that when there are many points in discussion periods which are not immediately considered for implementation, the psychological reward for the initiator is likely to disappear. The result is "just useless talking." As the physical laborers have somewhat less practice in manifold, delayed, and more abstract communications, the result is again somewhat dysfunctional.[4] Instead of increasing their motivation to identify themselves with the work problems, the appeal of the plant director seemed to build up defensive and rejective attitudes.

Along with the negative attitudes went an unusually high tendency to project responsibility for action upon other persons or groups. The functional analysis of organizations has shown that there was a considerable overlap of functions which stimulated the use of the projective defense mechanism. The resulting ambiguity in goals was due in a large degree to the organizational overlap. A reduction in the number of organizations would have probably brought about a more distinct delineation of particular functions.

There was one more point that should have been avoided.

---

[4] A similar point about the concreteness and controlled implementation of decisions was recorded by J. Woronowicz, "Niektóre Problemy Rad Robotniczych w Przemyśle Drobnym," *Drobna Wytwórczość*, VIII (January, 1958), 3-5.

The plant director sought to break up the unity between the workers and foremen by repeatedly asking the former to point out those foremen who did not care for their work. As group work and group discussion were simultaneously praised as the desirable way of behavior, the demand of the plant director sought actually to break up the solidarity of the work group. Furthermore, persons were aware of such a *divide et impera* strategy and retreated to ridicule and cynical statements.

Another point of dissatisfaction was the so-called swing shift. The reader will remember that the disturbing effect of the night shift was one of the few points on which both management and labor were agreed. Consequently, here was an opportunity to make a change which could have provided for some margin of improvement. Probably a more steady shift would have been a satisfactory solution.

On the whole, the management persons could have improved the communication between themselves and workers. In addition to barriers of distrust, workers actually looked upon a concrete problem, such as what are the causes of production difficulties, in a different way from the management. In terms of our model of workers' participation developed in the Introduction, the need was for more intensive and concrete participation in decisions. The practice of production discussions in the plant was extensive and general, while workers were more likely to get involved with just the opposite approach. Also, the management failed to realize that people are more likely to get themselves involved by doing something rather than being told what they ought to do.

Improvements on the organizational-institutional level were beyond the control of management or the workers' council. The Marxist theory, which influences the decisions of the party in Poland, assumes that there is no antagonism between management and labor. Władysław Gomułka, when he spoke at the IV Conference of the Labor Unions, defined the relationship in the following words: "Workers and management are not two antagonistic parts of the factory crew, but two sides of a unified organism, making possible its functioning."[5] A similar concept has been presented by Catholic encyclicals and other philosophies of social harmony and equilibrium.

The question we are now facing is as follows: To what degree can a norm influence, change, or even deny a phenomenon? Is it better, as it has been done in the West, to institutionalize the potentially antagonistic relationship between labor and management, or is it better to deny it and eventually to force the phenomenon out of the perceptual field by imposing, or trying to impose, a harmonious definition of the relationship between both groups?[6] Do we not get more conflict if we institutionalize it than if we deny it? Institutionalization might promote an antagonism even if there is no sufficient ground for it. On the other hand, a denial of differences, if not a total denial, might bring about a sudden eruption which might be quite disastrous to the survival of the social system.

The above problem has not been answered by our case study. The data we collected, however, suggest a tentative answer which can best be presented in terms of the structural-functional theory.[7]

We have already said that the appeals for group cooperation made by all organizations did not elicit the desired response. We would be inclined to conclude that workers were even

[5] Warsaw *Trybuna Ludu*, April 15, 1958.

[6] Our data seem to have enough validity to conclude that the relationship between management and labor in the factory was not harmonious and organically unified. The same observation probably was made in other Polish factories, because several authors living in Poland have referred to the antagonism. For example, see the demand that the party should mediate between the management and the labor, in Józef Balcarek and Maria Borowska, "Załoga a Rada Robotnicza," *Nowe Drogi*, XII (February, 1958), 65-81. Piotrkowski, 44, correctly pointed out that there was a conflict between the long-range plan and the interests of workers who did not identify themselves with the distant goals of economic development.

[7] The structural-functional theory, probably the most generally accepted theory among sociologists today, conceives of society as of a structure of more or less interrelated groups and institutions which contribute to the survival (eufunction) or hamper the survival (dysfunction) of society. See, e.g., Robert K. Merton, *Social Theory and Social Structure: Toward the Codification of Theory and Research* (Glencoe, Illinois, Free Press, 1949), 21-81.

more unwilling to cooperate genuinely as a result of this
direct pressure. In other words, the appeals had a dysfunctional
effect. Actually, though the workers sneeringly commented on
some official statements like "the workers rule the factory,"
they eventually used the same terms to justify and enhance
their own small spontaneous group actions. The official verbal
definition had some effect upon the rank and file—under certain
conditions. This effect cannot appropriately be described by
the term of dysfunction, because it helped the workers to
initiate certain positive changes and release tensions. These
changes were brought about, however, by means which were
not desired by the management. The management and the
party would have preferred that labor initiated action within
the organization that was designed to serve a somewhat similar
purpose. Thus, it seems to us that the best conceptualization
of this situation would be by the use of a new term—parafunction.

By definition, the parafunction is such a function which
stands between the eufunction and dysfunction. Since it uses
new uninstitutionalized means, it can disrupt the system to
whose survival it contributes. By acting on their own, workers
released a certain dissatisfaction but simultaneously created a
certain strain in their relationship with management. A good
illustration of this point is the attitude of the new manager,
who refused to accept the invitation of foremen to attend their
informal meeting for fear that they might become too demand-
ing. Because of the innovative indeterminateness, the parafunction can turn out both ways—as a survival function
(eufunction) as well as a dysfunction.

In the light of the above conceptualization, it seems that the
denial of the antagonistic relationship might bring about
anomalous emergencies within the organization. Thus we
propose that the denial of the differences between management
and workers, which is probably due more to the division of
labor than to ownership or nonownership, tends to develop
stress and tension within the organization. The institutionali-

zation of a certain antagonism, on the other hand, provides for a greater possibility of control and stability of the social system.

As we have already pointed out in Chapter V, the heart of the plant's labor-industrial relations was actually an inadequacy of the theory. The Marxist theory contains elements which are empirically untenable. The thesis about the disappearance of the conflicting differences between management and labor has failed to stand the empirical test. Not only that, but also the theory does not really explore the problem of "sharing of decisions."[8] The theory does not tell us in what way and to what degree a physical laborer is not only interested in the sharing of decisions, but also to what degree he is functionally able to do so if the production system has to operate successfully. While in the West several researches on this point have already been undertaken (see Introduction), in Communist countries no research of such a kind has been reported to my knowledge. Nevertheless, new organizations such as the workers' council have been introduced without first testing their potentialities and liabilities both experimentally and theoretically.[9]

Interestingly, regardless of all differences, both West and East seek to reduce the potential or explicit conflict between both major groups of production by postulating the desirability of harmonious cooperation between them. It will be a task of industrial sociology to uncover some new insights which would help the men of action to bring about the desirable changes with less waste in motivational energy and good will.

[8] Lenin made vague and even contradictory statements on this point. See Hammond, *Lenin on Trade Unions and Revolution: 1893-1917*, 82-84.

[9] Possibly this is also an example of the dysfunctional effect of the Communist theory which discards empirical research-minded sociology as a "bourgeois science." In Poland, sociology as an academic discipline was banned from universities during the "Stalin years," i.e., during 1949-1956. However, recently, a first empirical research on workers' councils has been reported from Yugoslavia. See Jakim Sinadinovski, "Prvi Pokušaji Empiriskih Istraživanja Radničkog Samoupravljanja," *Sociologija*, I, no. 1 (1959), 141-52.

Appendix

BYLAWS OF THE WORKERS' COUNCIL

1. In accordance with the law issued November 19, 1956 (The Diary of Laws of the Polish People's Republic, November 24, 1956, No. 53), a workers' council has been created in the cotton factory at Lodz

2. The workers' council will use a rectangular seal bearing the following inscription: "The Workers' Council of the Cotton Factory in Lodz."

3. The workers' council will operate the factory on behalf of all employees who perform according to legal standards, following the goal defined by the national economic plan. The workers' council will pass decisions within the framework of rules defined by these bylaws.

4. The workers' council consists of 113 members, elected by the employees by means of an internal electoral procedure. Before the expiration of its term, the workers' council will decide as to an eventual change in the number of the members of the council of the presidium.

5. The number of members of the workers' council may be changed as a result of expansion or introduction of new departments, or as a result of a reduction of the enterprise. The workers' council members are chosen from the employees of the factory. The term of office in the workers' council is for two years.

6. The workers' council will select from among its members a

presidium consisting of 15 persons, including the chairman and the vice chairman. Neither officer may simultaneously hold the position of plant director. As all production and other departments ought to be represented in the presidium, the representation is to be allotted as follows:

Medium Yarn Spinning Department 2
Rejects Spinning Department 2
Fine Yarn and Thread Spinning Department 2
Weaving Department 2
Finishing Department 2
Administration 2
Main Mechanical Section 1
Main Power Section 1
Transportation and Maintenance Sections 1

7. The plant director is a member ex officio of the presidium and of the workers' council.
8. The presidium of the workers' council is the executive of the workers' council and is responsible to the workers' council.
9. The presidium will employ a full-time, adequately qualified, paid secretary. The latter has a specialized function and is not a member either of the workers' council or of the presidium.
10. The workers' council members of the same department will constitute a departmental workers' council. They will elect from their members a chairman and a vice chairman of the departmental workers' council.
11. The workers' council may be dissolved before the expiration of its term by a vote of the general assembly of the factory's employees. A particular member of the workers' council may be removed from office by the employees of the department the member represents, provided that two-thirds of the members of the departmental workers' council presents it as a motion.
12. In case of removal or resignation of a member from a departmental workers' council, the workers' council will organize an election for a replacement member in the department of which the former member was a representative.
13. The members of the workers' council are entitled to be paid for time spent on those workers' council duties whose fulfillment reduces their regular income.

14. The function in the workers' council does not relieve the member of his working duties in the factory.

15. As long as a person is a member of the workers' council, he may not be discharged from his job, provided that he has not been removed from his workers' council office in accordance with section 11. The transfer of a workers' council member to a new permanent job within the factory against his will may be brought about only if the transfer is approved by the workers' council. A workers' council member may not be discharged from his job during the next two years after the expiration of his workers' council term, except according to stipulations listed under section 11.

16. All decisions by the workers' council and by the presidium must be made at their meetings. The decision is taken by a simple majority of votes and is valid if half of the workers' council quorum or half of the presidium's quorum is present, with the exception of topics listed under section 11. If there is a tie vote, the chairman's vote shall decide. A member of the workers' council may criticize the work of the plant director at workers' council meetings or at other assemblies of the employees, and at meetings of the presidium or of the departmental workers' council.

17. A general meeting of the workers' council will be held at least once in each quarter. The chairman of the presidium is obliged to summon a special general meeting of the workers' council upon the demand of one-third of the workers' council members, of the presidium, of the director, or of a departmental workers' council.

18. The announcement of a forthcoming meeting, with the agenda, shall be communicated to each member of the workers' council no later than five days before the date of the meeting.

19. Every employee of the factory is entitled to attend the workers' council meetings, provided that this does not interfere with his work duties.

20. The presidium's meeting will be held at least once each month. In addition, the chairman is obliged to summon the presidium to a meeting if demanded by any member of the presidium.

21. The workers' council is responsible to the employees of the factory and should report its activities once each quarter.

22. The workers' council may summon a meeting of all the employees concerning issues of a general nature if a referendum is required.
23. The functions of the workers' council are especially:
    1) To approve the yearly production plan of the factory.
    2) To approve the organization structure of the factory, and to approve the plan of technological development of the factory within set limits.
    3) To evaluate economic achievements of the factory as well as to analyze and approve the quarterly and yearly balance.
    4) To decide about the allotment of the profit appropriated by the factory.
    5) To approve the rules obligatory within the factory.
    6) To approve the system of work standards and basic work stipulations, as well as the standards on premiums.
    7) To approve the employment plan within the given limits of wages.
    8) To approve standards on dress of workers and protection of employees during work, as submitted by the labor union.
    9) To make decisions on the sideline production.
    10) To evaluate the work of the presidium and to give it new directives for future activities.
24. The decisions of the workers' council or the presidium on wages, on the plant fund, on housing, on social, health, and accident prevention problems, and on disciplinary rules are to be made in cooperation with the local labor union organization. Where there is disagreement between the labor union and the workers' council presidium, the matter should be decided at general meetings of the workers' council and labor union.
25. The workers' council may pass a decision by which a part of the workers, or all the workers of particular departments or sections, might be eliminated from sharing in the profit, if their irregular work has lowered the obtained balance. The sharing of the profit ought to be granted to those employees who have worked in the enterprise at least for half a year. Also entitled to sharing in the profit are those employees who have been transferred to the enterprise by higher authorities, provided that they have worked during the stipulated time limits. The workers' council might disregard this rule in certain exceptional cases.

26. The functions of the presidium of the workers' council are:
    1) To approve the operating plans of the factory.
    2) To approve basic measures dealing with improvement of production, especially the organization of work, technological processes, improvement of the quality and aesthetic properties of the products, increased productivity, improvement of accident prevention and health measures, and conservation of raw material and fuel.
    3) To decide on steps leading to improvement of discipline of work.
    4) To control and decide on capital investment in the factory.
    5) To decide on motions submitted by the director and departmental workers' council on the disposal of unused machines and tools.
    6) To consider financial and economic proposals.
    7) To analyze the monthly economic results of the factory.
    8) To approve contracts covering services, if the money is paid from the fund earmarked for services of nonemployees, provided that the charge is more than 5,000 zlotys.
27. The departmental workers' council is directly related only with the workers' council. Functions of the departmental workers' council have to be approved by the presidium of the workers' council, and they are:
    1) To develop operating production plans.
    2) To suggest changes in the organizational structure of the department.
    3) To develop a plan for the technological development of the department.
    4) To evaluate the economic results of the department.
    5) To suggest ways of disposing of unused machines and tools.
    6) To initiate motions concerning improvement of workers' attitude toward their work in the department.
    7) To divide the plant fund among workers of the department within limits set by the workers' council, while taking into consideration the viewpoint of the manager of the department.
    8) To develop suggestions on accident prevention and health improvement measures.
    9) The departmental workers' council is responsible to the employees of the department and gives them an account of

its activities once each quarter. The council is obliged to consider any suggestion which is handed in by any worker in the department.

28. The plant director and the vice directors are nominated and removed by the pertinent governmental agency with the approval of the workers' council. The workers' council is entitled to ask for the removal of the director and his deputies; the demand may be brought about upon a motion submitted either by the presidium or by the departmental workers' council.

29. 1) The plant director manages the factory and is responsible for its operations before the law as well as to the employees of the factory and to the superior governmental agencies.

    2) The functions of the director are especially:
      (a) to organize the production process;
      (b) to issue orders and suggestions related to the operations of the factory, following the plan or the resolutions made by the workers' council or by the presidium, as well as the instructions given by the supervising governmental agencies;
      (c) to represent the factory in its external relations, to conclude contracts and agreements with other physical as well as legal persons, to make certifications on behalf of the factory, while observing valid legal standards;
      (d) to decide personnel matters, while observing the stipulations on the rights of the workers' council members.

    3) The director is authorized to delegate a part of his functions to other employees of the factory.

30. 1) In the factory the principle of one-person directorship and one-person responsibility is valid.

    2) In matters of work relationships, the director is superior to all other employees in the factory.

31. No member of the workers' council is entitled to intervene in activities reserved only to the plant director.

32. If the presidium fails to make certain decisions within adequate time, the director is entitled to make decisions necessary for the fulfillment of the plan and continuation of the regular production. The director, however, is obliged to report on these decisions at the next meeting of the presidium in order to obtain formal approval.

33. The director is entitled to suspend the execution of a decision taken by the workers' council or by the presidium if he considers it contradictory to legal standards, economically disadvantageous, or in contradiction to the obligatory plan. In the case of a disagreement between the director and the workers' council, each of the parties may appeal to the agencies supervising the enterprise. If a further appeal from the decision of the agency is desired, any party may appeal to the Ministry of Light Industry.

34. The present bylaws are obligatory. Any amendment may be brought about by the workers' council with the approval of the employees.

35. The bylaws of the workers' council are approved by the employees.

Index

Absence rate, 63

Appeals: development of collective spirit, 64; dysfunctional effect, 65, 110-11, 136, 141, 144; demand to raise morale, 109, 111; solidarity of the work group, 142

Aspirations to move to other occupations, 38, 125

Attitudes toward work, 65, 122-25

Conflict between management and labor: xiii; line dividing labor from management, 80-82; difference concerning production troubles, 129-33; difference in outlook on major function of factory, 135

Concentration of power: effects, 101, 109, 138; "democratic centralism," 108; Workers' Council, 108

Cyrankiewicz, Józef, 68

Collective property: perception of, 116-19, 135

Data: collection, 9-11 passim; unused 11, 117; sources of, 11; questionnaire, 11, 18, 134

Foremen: duties, 22; Foreman I, 28, 40, 41, 46, 48, 61, 67, 71, 73, 78, 80, 86; foreman delegates to all-textile industry labor union meeting, 66; personal appeals, 79, 97; fired foreman and tram strike, 97, 99

German codetermination law of 1951, xviii

Gomułka, Władysław: 4, 8, 40, 44, 50, 58, 98, 102, 142; strike undermines, 71

Group: informal, 137; small, 138

Identification with factory and productivity, 120-21

I-5: 29, 46-52 passim, 84-86 passim, 91, 117, 122, 125, 137; reasons persons voted for, 37; to help section foreman, 76; workers rule in factory, 77; only workers are not afraid, 80, 86; expressing the voice of people, 83, 88, 91; complains of isolation, 84; distrusted by other workers, 87, 109; break through communication barrier, 93; failure to communicate to all persons, 140

Lodz: 13, 14; theaters or broadcasting stations, 68; 13th wage, 82; strike of tram employees, 113

Labor unions: 1, 2, 84; Polish labor movement, history, 1-5; Central Committee of Labor Unions, 1, 23, 50; political parties, 2, 3; principle of individual management, 3; Discipline Law in 1950, 3, 4; collective bargaining, 3; liberalization, 4; the party, 5; as highest body, 5; to draft new law on shop committees, 5; discredited, 5, 43, 107; trustee, 25; factory role, 50; election of labor union secretary, 51; social welfare function, 54; overlapping with social department of factory, 54; director, 107; overlapping with Workers' Council, 136

Manager of the Weaving Shop: iii, 31, 32, 41, 49, 55, 59; conflict with

Manager (*continued*):
plant director, 48, 62, 66, 69-70, 107; personality compared with plant director, 113

Management: content of messages to workers, 103; ambiguity of actions and goals, 105, 138, 139, 141; deflation of messages sent to workers, 136

Model of workers extensity-intensity participation: xv, 103, 104, 111, 139, 142

Marx: xiii, xiv; ambiguous definition of roles and goals, 105, 139; Marx-Leninist theory, 110

Natolin, 39, 71

October, 1956: 4, 39, 44, 52, 58, 140; Workers' Council, 5; before October, 55

Overtime, 25, 78, 79

Owen, Robert, xiv

Poznan: riots, 39; and Hungary, 40

Parafunction, 144

Party: forceful merger with the Polish Socialistic Party, 4; members, 26; Natolin, 39, 71; party candidates for Polish parliament, 40, 42; self-criticism, 95; ambiguous definition of role and goal, 105, 139; theory of integrative authority, 106; role defined, 108

Piłsudski, Józef, 1

Plan: production, 24-27, 59, 60, 63, 90; fulfillment of plan, 64, 128; qualitative plan, 64, 82

Plant: surplus of organizations, 111; surrounded by, 14-15; entry into, 16; first contact, 17-18

Plant director: 33-34, 40, 41, 59, 61, 88-89, 140, 142; investigating committee, 42; conflict with manager of weaving shop, 48, 62, 66, 69-70; production situation in textile industry, 63; fulfillment of plan, 64; verbal appeals, 65; criticized by departmental party secretary, 67; calming down workers, 85; projection of responsibility, 93-94, 106; group that wanted to get rid of, 98, 101; fired foreman, 99; pessimistic statement concerning workers' cooperation in management, 99;

Plant director (*continued*):
struggle for power, 101, 107; labor union, 107; personality, 108, 113; role endowed with stress, 113, 138

Poland, and the Soviet Union, 68

Polish: worker compared to Western worker, xiv; labor, history of, 1-5; Socialistic Party, and the Communist Party, 1, 2, 4; Ministry of Light Industry in Warsaw, 15; students, 41; election of January, 1957, 43; patriotism, 66; army, 68

Production troubles: 125; "downtime," 25; surplus of organization, 111, 128, 141; causes of, 126-32

Reasons why people are silent at meetings, 67, 93

Research problem: definition, 11; method, 8-11; role of researcher, 12; first contact, 17-18; research design, change of, 18; Polish colleague, 19; answers, 135-43; model of workers' participation, 142; parafunction, 144

Responsibility: projection, 93-94, 106, 141; self-defense, 110

Stalin years: 102, 140; de-Stalinist process, 39; former Stalinist practices, 140

Saint-Simon, xiii, xiv

Section foreman: 23, 30, 46, 59, 74-77; personal appeal for overtime work, 77; and workers, 81, 140

Strike of tram employees, 113

Studies on workers' participation in management, xv-xviii; on work satisfaction, xv

Weaving shop (work groups I & II): description of, 19-22; shop time, schedule of shifts, 22; wages, 23, 24, 25; production plan, 24-27; overtime, 25, 78, 79; the party, 26, 33; transmission of news, 28; manager of, 31-32, 41, 44; workers' education, 34; war experiences of workers, 35, 113; average age, 35; marital status, 36; apartments of workers, 36; expenditure, 36; pastime activities, 37; occupational aspirations, 38; dividing of premiums, 49

Workers' Council: shop committees, 4, 5, 50; bill, 6-8; plant fund bill,

Workers' Council (*continued*):
8; local rules of, 44-45; 13th wage, 65, 82-96, 104; evaluation, 87; rights of departmental council, 90; concentration of power of, 108; overlapping with labor union, 128, 136

Workers in plant: characteristics in weaving shop, 36-38, 113; aspirations to move to other occupations, 38, 125; collective action, 55-56, 72-73, 76, 92, 98, 112, 139, 141; complaints by, 60-62; absence

Workers in plant (*continued*):
rate, 63; failure to speak out, 67, 93; distrust of management, 80, 95, 109, 136; lack of interest in management, 87; "rule the factory," 102, 111, 144; disrespect toward labor union, 107; perception of collective property, 116-19; identification with the factory and the productivity, 120-21; feeling about work, 122-25

Yugoslav Workers' Council, 6, 7, 8